S0-BSU-574

Concerned About the Planet

Recent Titles in
Contributions in American Studies
SERIES EDITOR: ROBERT H. WALKER

Sodoms in Eden: The City in American Fiction before 1860
Janis P. Stout

Appointment at Armageddon: Muckraking and Progressivism in American Life
Louis Filler

Growth in America
Chester L. Cooper, editor

American Studies Abroad
Robert H. Walker, editor and compiler

The Modern Corporate State: Private Governments and the American Constitution
Arthur Selwyn Miller

American Studies: Topics and Sources
Robert H. Walker, editor and compiler

In the Driver's Seat: The Automobile in American Literature and Popular Culture
Cynthia Golomb Dettelbach

The United States in Norwegian History
Sigmund Skard

Milestones in American Literary History
Robert E. Spiller

A Divided People
Kenneth S. Lynn

The American Dream in the Great Depression
Charles R. Hearn

Several More Lives to Live: Thoreau's Political Reputation in America
Michael Meyer

The Indians and Their Captives
James Levernier and Hennig Cohen, editors and compilers

Martin K. Doudna

Concerned About the Planet

THE REPORTER Magazine and American Liberalism, 1949-1968

Contributions in American Studies, Number 32

GREENWOOD PRESS
WESTPORT, CONNECTICUT · LONDON, ENGLAND

Library of Congress Cataloging in Publication Data

Doudna, Martin K.
Concerned about the planet.

(Contributions in American studies; no. 32)
Bibliography: p.
Includes index.
1. Liberalism--United States--History. 2. United States--Politics and government--1945- 3. United States--Foreign relations--1945- 4. The Reporter (New York, 1949-) 5. Press and politics--United States. I. Title
E743.D66 320.5'13 77-10048
ISBN 0-8371-9698-1

Library of Congress Catalog Card Number: 77-10048
ISBN: 0-8371-9698-1
ISSN: 0084-9227

First published in 1977

Greenwood Press, Inc.
51 Riverside Avenue, Westport, Connecticut 06880
Printed in the United States of America

For Dorothy

Contents

Preface

This book began with my curiosity about why a prominent liberal magazine, *The Reporter*, suddenly ceased publication in June 1968. When it first appeared in 1949, I was a freshman at Oberlin College, and I recall buying a copy of one of the famous China Lobby issues in the spring of my senior year. I was delighted a few months later when it switched from Eisenhower to endorse one of my first political heroes, Adlai Stevenson. My sustained interest in the magazine, however, did not begin until about ten years later when I was a civil servant in Washington, D.C., reading as part of my job a vast array of newspapers and magazines every week. Among them *The Reporter* proved to be one of the best single sources of information and ideas.

In 1949, when *The Reporter* began publication, postwar liberalism in America was in its heyday: it was the popular thing, at least among intellectuals, to identify oneself as a liberal. But when it ceased publication nineteen years later, liberalism had come to be considered suspect or out of date. Phrases like "the poverty of liberalism," "the end of liberalism," and "the death of liberalism," all of which appeared as titles of books or articles, had become almost clichés. The end of the most widely read American magazine that explicitly identified itself as "liberal" seemed to coincide, in a neat and almost uncanny fashion, with the reputed end of liberal dominance in American thought.

But as I quickly discovered, things were not that simple. While the magazine's contributors and readers represented a cross section of American liberalism, its founder, publisher, and editor, Max Ascoli, came out of a different tradition, that of European liberalism. And if *The Reporter* was in many ways a reflection of American liberalism, in other ways it represented a deviation from it. It opened its pages to a variety of liberal spokesmen; at the same time it continually criticized a number of liberal clichés. And ironically, some of the shortcomings of liberalism that contributed to the crisis of liberalism at the end of the 1960s were the very clichés most persistently criticized by *The Reporter*.

The book opens with a discussion of postwar American liberalism, a discussion that attempts to define that elusive term, to indicate briefly what I regard as one of its major weaknesses as a world view, and to show the relationship between liberal magazines and the liberal community. The literary allusions that appear in the first chapter, and subsequently, reflect the fact that I am not a political scientist but a professor of English literature: I find it artificial to talk about liberalism without talking about liberals, and I think man is too complex, elusive, and unexpected a creature to be explained solely within the frames of reference devised by philosophers or by social scientists.

The second chapter is concerned with the background and thinking of that most fascinating and too-little-known man, Max Ascoli, and the third deals with his associates and with the early history of *The Reporter*, together with facts about its format, contents, and circulation. The fourth and fifth chapters deal with *The Reporter*'s views on domestic policy and on foreign policy respectively, and the sixth chapter completes the history of the magazine by telling the story of its closing in the spring of 1968.

A final chapter presents some conclusions and discusses briefly four fallacies in liberalism that I see as emerging from the study. I am aware that the final chapter, like the first chapter, raises some large questions. I doubt whether I have answered them to anyone's complete satisfaction, including my own; but I hope that raising them will encourage others to join in the search for the answers, since the ultimate value of any study—like the ultimate value of *The Reporter* and of liberalism itself—lies not so much in itself as in what it may inspire others to think and to do.

For the chapters dealing with *The Reporter* itself, I am first of all deeply indebted to Max Ascoli, who granted me an extended interview in January 1971 and commented carefully on the factual accuracy of my references to him and his magazine in the first draft, and who has generously given me permission to quote from *The Reporter* and from his writings elsewhere. I might add that he is not in agreement with all of my interpretations; it would hardly be in character if he were. Two members of his staff—Elizabeth Parsons, who set up the interview, and Ruth Ames, who gave me access to her files as librarian and sent me an account of the magazine's prepublication days—were particularly helpful.

I am likewise grateful to the former staff members—Donald Allan, George Bailey, Robert Bendiner, Robert Bingham, Harlan Cleveland, Philip Horton, Shirley Katzander, Irving Kristol, and Derek Morgan—whom I was able to interview. Among them I found strong—and sometimes contradictory—opinions, but they were all patient with the queries of a stranger, and I owe much to the information and insight they provided.

I owe a special debt of gratitude to Joe Lee Davis, John Higham, Robert Haugh, and Ben Yablonky for their encouragement, advice, criticism, questions, and friendship when I was writing the first version of this book at the University of Michigan, and I am grateful to the University of Michigan and the Ford Foundation for a Horace H. Rackham Fellowship which provided financial support at that time. Helpful comments on the original manuscript were made by Edric Cane, Bruce Currie, and Morris Martin. John Daeley's knowledge of Italian proved valuable, as did the practical suggestions I received from Michael Meyer. My cousin Phyllis Doudna and my neighbor Kate Bonsey typed successive drafts with great efficiency and cheerfulness. And finally, for her help with the proofreading, for her incisive comments that improved the style and accuracy of the text, and for her continual encouragement, I want to thank my wife Dorothy.

Martin K. Doudna
Hilo, Hawaii
January 1977

Chapter One

Liberalism in Postwar America

In a sense all of America is liberalism.

Arthur M. Schlesinger, Jr., 1956

The American dream was killed by liberals.

Senior at Columbia University, 1970

Liberalism in America, at the end of the 1960s, found itself in trouble—trouble that might well be described in words John Dewey had used in the mid-1930s:

> Liberalism has long been accustomed to onslaughts proceeding from those who oppose social change. It has long been treated as an enemy by those who wish to maintain the *status quo*. But today those attacks are mild in comparison with indictments proceeding from those who want drastic social changes effected in the twinkling of an eye, and who believe that violent overthrow of existing institutions is the right method of effecting the required changes.[1]

The appeal of the left to American liberals in the 1930s is well-known. Alfred Kazin, who was a student in the middle of that decade, recalls that in 1937, even though he and his friends distrusted Stalin and disbelieved in the Moscow Trials, all the "cleverest and most dynamic people

I met . . . gave authority to Marxist opinion."[2] During World War II the common objective of defeating Nazi Germany joined the United States and the Soviet Union in an alliance that seemed to many to transcend ideological differences. Criticism of the Soviet Union was often muted or replaced by praise, and Communists were accepted as agents in the highly sensitive Office of Strategic Services.[3]

After the war many Americans hoped to continue good relations with the Soviet Union, despite a growing suspicion that the gulf between democracy and dictatorship was too wide to be bridged by good intentions. There was honest confusion, as well as some deliberate conniving, in liberal circles. On one side were those whom Dwight Macdonald scathingly denounced in 1945 as "totalitarian liberals." Macdonald accused these liberals of advocating antiliberal policies to arrive at "practical" solutions, of employing a double standard of political morality in judging actions by the United States and actions by the Soviet Union, of regarding the Soviet Union as "the repository of all political virtue," and of believing that society is the end toward which individual human beings were only the means. In short, although they called themselves liberal, their beliefs were totalitarian. Macdonald drew his examples of totalitarian liberalism from the *New Republic*; he claimed that he could have taken them as easily from the *Nation*, which he characterized as differing from the *New Republic* only in being "more fuzzy minded, naive, and idealistic."[4]

Such "totalitarian liberals" formed the leading elements in the Progressive Citizens of America, a group which was organized in 1947 from two former Communist fronts and which became the core of Henry Wallace's Progressive Party. Opposing them in the battle to represent American liberalism were the 250 liberals—led by such figures as James Loeb, Jr., Reinhold Niebuhr, Chester Bowles, Arthur Schlesinger, Jr., Walter Reuther, Eleanor Roosevelt, Joseph Rauh, Marquis Childs, David Dubinsky, Leon Henderson, Wilson Wyatt, Barry Bingham, and James Wechsler—who organized the Americans for Democratic Action early in 1947.[5]

Explicitly anticommunist from the beginning, the ADA successfully fought Wallace's third party, and by the fall of 1948 ADA members found themselves, somewhat to their embarrassment, supporters of a reelected Harry Truman, a man for whom they had felt only a tempered enthusiasm. Truman's second term was the high point of what was then called the "new liberalism."[6] "In the United States at this time liberalism is not only the dominant but even the sole intellectual tradition," Lionel Trilling confidently affirmed in 1950.[7] Trilling's claim appeared to be substantiated when Louis Hartz's influential "consensus" interpretation of the history of American political thought, *The Liberal Tradition in America*, was published in 1955. For the rest of the fifties the ADA represented a righteous opposition to President Eisenhower, the man whom, ironically, ADA leaders had proposed to replace Truman as Democratic candidate in 1948 when they thought Truman could not win. During this period the ADA was probably "the most representative liberal group operating on the national scene."[8]

John F. Kennedy did not stir much more enthusiasm in ADA breasts than had Harry Truman. Kennedy had told an interviewer in 1953, "I never joined the Americans for Democratic Action. I'm not comfortable with those people."[9] Arthur Schlesinger, Jr., arguing strenuously for Kennedy's election in 1960, described him as "a committed liberal," but still had to concede, "He is not particularly committed by spontaneous visceral reactions in the usual pattern of American liberals."[10] And in a poll of a hundred college and university professors taken by *Esquire* in late 1959, only two—John Kenneth Galbraith and Crane Brinton—had supported Kennedy.[11] But when it came to a choice in the 1960 election, the duty of the liberal seemed clear: Kennedy might be young and inexperienced, he might be a Catholic, and some of his past political positions might have been unfortunate, but the alternative—Richard Nixon in the White House—was unthinkable.

When Kennedy entered the White House, he brought a number of ADA members with him. Schlesinger, a leading ADA spokesman, joined the White House staff, and between thirty and forty other members took key positions. Things were looking up for liberals; the only major flaw of the New Frontier was the Vice President, a man whose voting record had been rated by the ADA as a trifle more liberal than that of the President, but whom, nevertheless, many ADA liberals could not abide.[12] After his succession to the Presidency, Lyndon Johnson could point in vain to his impressive legislative achievements; he simply did not have the "style" of a Kennedy nor even of his own political hero, Franklin Roosevelt.[13] The Vietnam War, which Johnson had inherited from his predecessors, grew during his administration into a major disaster. The civil-rights legislation pushed through by Johnson was not enough to prevent an eruption of racially related riots in the urban ghettos in the summer of 1967. In 1968 came a series of shattering events: President Johnson's decision not to seek reelection, announced at the end of March; the murder of Dr. Martin Luther King, Jr., a few days later; the murder of Senator Robert Kennedy in June; the street disturbances during the Democratic Convention in Chicago; and in November the election of Richard Nixon, the one political leader most consistently disliked by liberals for nearly twenty years.

Twenty years earlier, Schlesinger had bravely proclaimed that liberalism was democracy's "vital center."[14] Now there seemed to have been an uncomfortable prophecy in the words of Yeats with which Schlesinger had prefaced his book: "Things fall apart; the centre cannot hold." In the spring of 1970 W. H. Auden noted casually in a book review that he and his readers were living "in an age when *liberal* has become a dirty word."[15]

The vicissitudes of postwar American liberalism cannot be explained wholly in terms of the impact of events and personalities. If character is fate, then the plight of liberalism at the end of the 1960s can be understood only in light of its own nature and complexities. Some of these will emerge as we turn successively to a consideration of the liberal ideology, the liberal temperament, criticisms of liberalism made by outsiders, and the liberal community.

THE LIBERAL IDEOLOGY

Liberalism, as anyone who has tried to define it will readily admit, "is not easy to describe, much less to define, for it is hardly less a habit of mind than a body of doctrine."[16] Lionel Trilling speaks of "the ideas of what we loosely call liberalism," which he characterizes as "a large tendency rather than a concise body of doctrine."[17] Another writer has noted:

> In contemporary usage the term "liberalism" refers to a system of thought and practice that is less specific than a philosophical doctrine, and more inclusive than a party principle. Liberalism is also too ecumenical and too pluralistic to be called, properly, an ideology.[18]

Fiorello LaGuardia once complained that liberals, "like soup, came in fifty-seven varieties,"[19] and Thurman Arnold noted in 1946, "Liberalism today has become deuces wild. It can be used to fill any hand."[20] Adlai

Stevenson, addressing the Liberal Party of New York in 1956, summed up the problem of definition adroitly: "No man can precisely define liberalism without getting into an argument with another liberal; that is the nature of our creed—and breed."[21]

Periodically there have been attempts to reduce liberalism to a kind of dogma:

> The word "liberal" as used by the New Dealers about themselves began to take on specific meaning. At first it meant only the search for answers to the problems of economic depression. Later it came to involve loyalty to the aspirations of organized labor and racial minority groups. In order to qualify as an American Liberal a man had to favor compulsory FEPC, regard Taft-Hartley as completely evil, believe in the total wickedness of Chiang Kai-shek, take his political economics from John Maynard Keynes, and look to Washington for the answers to all problems. The movement became more cohesive. It gave itself organizational form in Americans for Democratic Action. It evolved not only its theology but also its hierarchy of saints and devils. A political movement born of the depression had become a political religion.[22]

But as the same writer goes on to note, "True liberalism is not a dogma, it is a point of view."[23] And this point of view, for at least three reasons, tends to include a wide spectrum of political beliefs.

First, both "liberal" and "conservative" are relative terms. In 1952 Eisenhower may have been conservative as compared with Adlai Stevenson, but he was liberal compared with Robert Taft. Furthermore, the terms are relative with respect to time. Reforms that appear liberal in one period are regarded as conservative a few years later. Thus, as Eric Goldman noted in 1953, "In recent years, practically everyone, or at least practically everyone seeking national approbation in the United States, has become something of a liberal in the old Rooseveltian sense."[24] In the 1960 election, Amitai Etzioni pointed out, both platforms were relatively liberal: the entire political climate was moving toward a liberal orientation.[25]

Second, the word "liberal" has an emotional resonance that leads many to find personal satisfaction or political advantage in so describing themselves—or in so describing their opponents. Sometimes a qualifying adjective is added. Thus we have "conservative liberals," "realistic liberals," "tough-minded liberals," "critical liberals," and so forth. Some of the qualifying adjectives are pejorative, as with Macdonald's "totalitarian liberals," Granville Hicks' and Sidney Hook's "ritualistic liberals," and Robert Bendiner's "automatic liberals," who all turn out to be cousins of that creature so detested by conservatives, the "knee-jerk liberal." Sometimes a qualifying adjective can be favorable or unfavorable, depending on who is using it, as with the term "radical liberal," which was apparently invented by the political scientist Arnold Kaufman and used favorably by him, but which was widely popularized by Vice President Spiro Agnew as a term of derision.[26]

When the *New York Herald Tribune* posed that perennial query, "What is a Liberal?" in 1949, one reader responded, "Practically everybody I know is a 'liberal.' At least that's what they call themselves."[27] In the same year Robert Bendiner wrote, "Out of some 140,000,000 people in the United States, at least 135,000,000 are liberals, to hear them tell it, liberal having become a rough synonym for *virtuous*, *decent*, *humane*, and *kind to animals*."[28] It is hardly surprising that as a result the word "liberal," for some people, began to lose much of its emotional force. Toward the end of the 1960s it had come to sound increasingly timid, soft, fuzzy, vague, and tentative. "Radical" was rediscovered as a word with the hard, bold, decisive sound that an age of crisis seemed to demand. With a certain sense of déjà vu one can compare Arnold Kaufman's statement in 1968, "Confronted by the sordid reality of American affluence, it is impossible for someone to be authentically liberal without turning resolutely toward radicalism," with John Dewey's statement of more than thirty years earlier: "If radicalism be defined as perception of need for radical change, then today any liberalism which is not also radicalism is irrelevant and doomed."[29]

Finally, as a third factor that contributes to semantic and ideological confusion, many liberals think liberalism can best be defined as a "faith" or "spirit" that must manifest itself quite differently in different historical circumstances. The definition written in part by James Wechsler for the organizational meeting of the ADA in 1947 reflects this view:

> Liberalism is a demanding faith. It rests neither on a set of dogmas nor on a blueprint, but is rather a spirit which each generation of

> liberals must learn to apply to the needs of its own time. The spirit itself is unchanging—a deep belief in the dignity of man and an awareness of human frailty, a faith in human reason and the power of free inquiry, a high sense of individual responsibility for oneself and one's neighbor, a conviction that the best society is a brotherhood that enables the great numbers of its members to develop their potentialities to the utmost.[30]

David Lilienthal, who as a New Deal liberal and director of the Tennessee Valley Authority had been distrusted by businessmen until he left government service to become a successful businessman himself, has argued that liberalism "has nothing to do with political or economic beliefs." Its basic concern, rather, is with individual freedom. Government intervention, he says, may be necessary to guarantee that freedom against "the brutal, the ruthless, the covetous side of man's nature." But government action in other instances—under a Communist or Fascist dictatorship, for example —may be inimical to freedom. So liberalism, Lilienthal concludes, comes down to being "a basic faith in freedom as the important thing."[31] Robert Bendiner once put it more succinctly: "For my money, a man is a liberal as long as he tries actively to make the underprivileged the privileged and to check them if and when they become overprivileged."[32]

This conception of liberalism as a faith is attractive, but it opens the liberal to charges of inconsistency: as a defender of business, Lilienthal encountered strong reactions from some of his New Deal friends.[33] It also means that liberalism tends to define itself in negative terms, as "the urge to wage war against injustice, against tyranny, against selfishness, against inhumanity from whatever direction those qualities may come."[34] Jefferson's ringing words, "I have sworn on the altar of God eternal hostility against every form of tyranny over the mind of man," make every true liberal's heart beat faster. But if he is not careful, the liberal is likely to feel more at home when arrayed against such foes as Southern segregationists, the Republican old guard, the military-industrial complex, or totalitarian Communism than when he is proposing a positive program of his own. Having espoused, in effect, an ideology of "no ideology,"[35] liberals include among their ranks anticommunists, anti-anticommunists, and even a few anti-anti-anticommunists. The liberal—and he is not alone in this—glories in his enemies. Sometimes one feels that if they did not exist, he might almost be tempted to invent them.

When asked what he is *for*, the liberal sometimes slips into an embarrassingly vague rhetoric. As a result, liberalism may face its greatest danger at the moment of its apparent triumph—when it achieves political power. It is hard to deny some truth to Michael Harrington's assertion that when Lyndon Johnson's overwhelming electoral victory in 1964 seemed to make the traditional liberal program the official ideology of the United States, "liberalism, as it was known for a generation, died."[36] In carrying out their own political programs, liberals find that compromise is often necessary. A liberal administration finds that it must make decisions—and there are always tough ones—which have consequences to be endured and defended. These seem to be the inevitable conditions of political life in a democracy; but political theorists—often liberals themselves—sometimes grow impatient and long for the governmental efficiency that liberal democracy frequently does not seem to allow.[37]

Political theorists sometimes appear to forget an obvious truth: the liberal is not only an abstraction, he is also a human being. He is far more likely to be out working for practical solutions to real problems than to be meditating in his study on pure political truth. Liberalism is always something of an ad hoc philosophy, shaped by the pressures of its time. The liberal as an abstraction can be labeled, classified, and defined; as a human being he is constantly resisting labels and breaking out of any categories into which he is placed. Unpredictable and inconsistent, he is the child of his times. He may simply "move with the times" as do many politicians, or he may, like some intellectuals, be a compulsive dissenter—even from dissent itself. Liberals like to think of themselves as a special breed. They share, however, the common human tendencies to see most issues as an inevitable choice between two alternatives, to overstate their own case, to make tactical retreats and strategic blunders, and to act on the belief that "the friend of my enemy is my enemy, and the enemy of my enemy is my friend." These characteristics have to be kept in mind as we look at attempts to locate liberalism on the political spectrum.

The political spectrum is frequently pictured as a line extending from reactionaries on the extreme right to conservatives, then liberals, and finally radicals on the extreme left. Though oversimplified, this picture shows graphically how a certain foreshortening may occur. Those on the right side of the spectrum can see little distinction between liberals and radicals; those on the left have similar difficulty in distinguishing conservatives and reactionaries. The further one stands toward either extreme, the more pronounced is this tendency.

Arthur Schlesinger suggested in 1949 that this linear diagram be replaced by a circular one that would show the similarities of the extremes on both the left and the right. Since Communism and Fascism are both totalitarian ideologies, committed to violence and hostile to freedom, those between the two extremes are natural allies against them. Although Schlesinger tried to maintain a distinction between liberals and conservatives, together they tended to make up what he called the "vital center," a conception ideally suited to Schlesinger's position at the time as one of the spokesmen for the anticommunist ADA.[38]

In both the linear and circular diagrams, the liberal, unfortunately, ends up somewhere in the middle. Frank Sullivan's description overstates the case—"a liberal is a man who is constantly and simultaneously being kicked in the teeth by the Commies and in the pants by the National Association of Manufacturers"[39]—but nevertheless the liberal is often uncomfortably aware that he must wage battle on two fronts at once. And the enemy on each side is at the same time a potential ally against the other. Being the man in the middle has another disadvantage. Where extremists at either end of the spectrum can stir their followers by visions of sweeping, apocalyptic solutions to all social problems, the liberal is too often in the unexciting position of saying, "Yes, but . . . ."

The defects of theoretical definitions of liberalism lend strength to Wayne Morse's argument: "Liberalism cannot be defined in the abstract in any helpful way. Liberalism in politics can best be defined in terms of specific issues."[40] Liberals in the postwar period, of course, have been generally in favor of civil rights, civil liberties, social welfare, labor, foreign aid, and the United Nations. But liberal positions on some of these issues may change with time or circumstances. United States foreign policy is a crucial issue on which many liberals have modified or even reversed their positions in the past thirty years.

In the aftermath of World War II many liberals were internationalists. The six-point statement of principles released after the founding of the ADA in January 1947 ended with these two:

> Point 5: Because the interests of the United States are the interests of free men everywhere, America must furnish political and eco-

nomic support to democratic and freedom-loving peoples the world over.

Point 6: Within the general framework of present American foreign policy steps must be taken to raise standards of living and support civil and political freedoms everywhere.[41]

These principles were consonant with the Truman Doctrine announced by the President two months later, and ADA liberals supported both the Truman Doctrine and the Marshall Plan. When Truman announced American armed support for South Korea three years later, most liberals, like most conservatives, supported him.[42]

"We are all liberals now," Chester Bowles claimed cheerfully in 1959,[43] and in 1966 John Kenneth Galbraith argued that for the past twenty-five years "to a remarkable extent . . . liberals and conservatives have pursued the same objectives in both foreign and domestic policy." In foreign policy, he maintained, "the goal to a singular degree has been to fight Communism," for "liberals long ago accepted simple anti-Communism as a goal of our foreign policy."[44]

If this view was correct, a disturbing corollary followed: Vietnam was "a liberal's war."[45] But many liberals would not accept this conclusion. The twenty-one "prominent intellectuals of liberal or democratic-socialist persuasion" who took part in *Commentary's* symposium, "Liberal Anti-Communism Revisited," in 1967, all opposed the Vietnam War, but only a few of them felt that liberal anticommunism could be held responsible for it. Richard Rovere, one of this small minority, maintained that the ADA, "to which I have never belonged," had by its militant anticommunism helped create a climate favorable to the war, adding, "Dean Rusk's speeches put one very much in mind of the kind of thing one heard at rallies of the American Committee for Cultural Freedom fifteen years ago—or, for that matter, the kind of thing one read at the time in the columns of this magazine."[46] But Michael Harrington strongly disagreed, saying that it was "McCarthyism" to blame Vietnam on the liberals, since a majority of those who opposed the war did so from liberal, internationalist motives.[47] Sidney Hook, an unrepentant anticommunist, argued that "the anti-Communist liberal is committed to a double strategy—a program of social change or reform under the protective umbrella of a military defense that guarantees the freedom to initiate protective measures"; and he attacked the "ritualistic liberals" who think the only defense against totalitarianism is social reform: "If we had accepted their strategy against Hitler most of us would be dead by now."[48]

A number of New Frontiersmen, equally reluctant to see liberalism as the cause of the Vietnam War, came up with an easy explanation for it—the pugnacious egotism of Lyndon Johnson. "If John Kennedy had lived . . ." they would begin, implying that he would surely have found some way out of the morass. Since their evidence consisted largely of unsupported personal recollections or glib readings of Kennedy's character ("too smart an Irish Mick"), their assertions were not very convincing.[49]

Other liberals did not agree that the issue could be reduced to the personality of the man in the White House. Rather, many liberals, like many radicals, became skeptical about the whole basis of liberal foreign policy.[50] Foreign aid, a liberal shibboleth in the 1950s, increasingly came under liberal attack. American involvement in Vietnam, it was argued, began in innocence with economic aid. As economic aid came to be accompanied by growing military assistance, the United States found itself, willy-nilly, deeply involved in a land war in Asia. Since the links between economic assistance and military assistance seemed so close, many liberals began to doubt if Uncle Sam could act as global Good Samaritan without acting as global policeman as well. For they had seen the cherished liberal dream of extending the New Deal worldwide—of giving, in Henry Wallace's words, a quart of milk to every child in the world—grotesquely transformed into a Vietnam nightmare.

It is not surprising, then, that disillusioned liberals of the late 1960s, like disillusioned liberals after World War I,[51] should have turned their backs on their previous idealism and have succumbed to a somewhat cynical isolationism, This "neo-isolationism"—most liberals indignantly rejected the term—expressed itself positively in a call for domestic reforms. As Eric Goldman wrote in 1969: "In the many rooms of the rambling house of American liberalism, one has frequently been occupied by a muzzy isolationism, and this fact now came to the fore in the agitation conducted by a number of the reform-minded."[52]

Thus foreign policy illustrates the dangers of trying to define liberalism merely in terms of its stand on specific issues. It must somehow consist of something deeper. One critic of liberalism, Robert Paul Wolff, has maintained that the deeper element is a philosophical world view—an inadequate one, he argues—derived from John Stuart Mill.[53] Wolff's demonstration of the flaws in Mill is convincing, but his tacit assumption that modern liberalism derives simply from Mill's philosophy is questionable. If William James is correct in his assertion that the history of philosophy "is to a great extent that of a certain clash of human temperaments,"[54] then liberalism may have its roots as much in the liberal's temperament as in his philosophy.

THE LIBERAL TEMPERAMENT

Liberalism has been defined by one liberal as "fundamentally . . . an attitude. The chief characteristics of that attitude are human sympathy, a receptivity to change, and a scientific willingness to follow reason rather than faith or any fixed set of ideas."[55] Another notes that "liberalism is to a significant degree a movement of conscience and tends to attract a particular type of personality."[56] The conception of the liberal as a type of personality may help explain such apparent ideological inconsistencies as the persistence of Thomas Jefferson as patron saint of modern liberals. Over the years Jefferson has been virtually all things to all men; his positions on the power of government, economic policy, slavery, and the press have been cited by liberals and conservatives alike.[57] Herbert Croly, an early theorist of welfare liberalism, reacted against Jeffersonian individualism and in his book *The Promise of American Life* (1909) argued for a Hamiltonian reliance on a strong national government. Trying to get the best of both worlds—a characteristic trait of the liberal temperament—Croly urged the "pursuit of Jeffersonian ends by Hamiltonian means."[58] Forty years later, Arthur Schlesinger, Jr., echoed this appreciation of a Hamiltonian strong federal government.[59] Thus twentieth-century American liberals proclaim themselves Jeffersonians while acting like Hamiltonians, an anomaly best explained by the assumption that much of Jefferson's real attraction for liberals is not as an ideologue but as a personality. No matter what embarrassing Jeffersonian quotations the conservatives may dredge up, Jefferson simply remains the kind of man with whom the liberal can identify.

Frederick Howe, the Progressive reformer, has given a good picture of the liberal personality:

> It was good form to be a liberal; it involved no sacrifices. Indeed, it gave distinction. I never lost my feeling of being one of the elect,

> of helping keep America true to the ideals of the fathers. I believed that the things I wanted would come about in time; that they would be brought about by liberals—liberals as represented by the New York *Nation*, the *New Republic*, the insurgent group in Congress. I was confident that peace societies would end war. I believed in discussion; in the writing of books and magazine articles, in making speeches. We liberals had the truth. If we talked it enough and wrote it enough, it would undoubtedly prevail. By eloquence and reason abuses would be ended; the state would be cleaned up. I believed in the mind and in facts.[60]

Most lists of the characteristics of the liberal personality would probably begin with openmindedness—a profound belief in freedom of thought and discussion. It has been suggested that the basic definition of a liberal is "a man who defends the personal liberty and freedom of thought of those whose opinions he scorns."[61] It is always easier, however, to defend freedom of thought and expression in the abstract than in real situations where such freedom appears dangerous. A basic liberal dilemma appears in Milton's classic plea for freedom of the press, "Areopagitica": "Popery and open superstition, which . . . extirpates all religions and civil supremacies," and "that also which is impious or evil absolutely either against faith or manners no law can possibly permit, that intends not to unlaw itself." In other words, civil liberty that allows the extinction of civil liberty is a contradiction in terms. In the 1950s the enemy was not "popery and open superstition" but Communism; the problem, however, was the same: how much toleration could safely be given to those committed to the destruction of toleration? The formula enunciated by Sidney Hook in his book *Heresy, Yes; Conspiracy, No* (1953) was one that could be accepted by many liberals. But the line between the two was not always easy to draw. In the era of Joe McCarthy, civil liberties became a heated issue for both liberals and their opponents.

A second characteristic of the liberal is his conscience. Emotionally he is on the side of the underdog; he works on behalf of ethnic minorities and the poor. He tends to be progovernment and antibusiness, and he advocates greater spending for programs of social welfare. He likes to think of himself as a reformer.

Along with this goes a belief in progress and in reason as an instrument of progress—a view that pervades the writings of John Dewey. Sidney

Hook, a student of Dewey's, defines a liberal as "a person committed to the use of intelligence in behalf of the values of freedom."[62] This liberal faith in the use of reason squares with the fact that historically American intellectuals—academics, journalists, librarians, and scientists—have tended to be liberals.[63]

Since he is committed to the use of reason, the liberal is normally committed as well to science and technology, which employ reason in an organized fashion to gain new knowledge and then use that knowledge to help solve the practical problems of housing, health, sanitation, and nutrition. But by the mid-1960s a reaction against science and technology had begun to take place among the young. This youth "counter culture" opposed not only science and technology but also the liberalism associated with them.[64]

With his belief in reason, science, and progress, the liberal tends to be skeptical about religion and about tradition in general. He is not so much antireligious as irreligious—religious questions for him are simply not "live" issues. If pressed, he may admit that many of his values may ultimately have had a religious origin, but since God so often seems to have been taken over by the conservatives as their exclusive property, the liberal will merely say that he is interested not in God, but in man. In his concern with what he regards as immediate and practical questions, the liberal tends to concur with Jefferson's sentiment, "The earth belongs to the living."

Sometimes it seems that it is almost a definition of the liberal to say that he—like John Henry Newman's famous "gentleman"—is one who wants to make everybody happy. It is in determining how to attain this desirable end that the greatest split in liberalism occurs. On one side are the "economic" liberals—those who support welfare state measures, higher wages, graduated income taxes, and labor unions; on the other are "noneconomic" liberals—those who support civil liberties, civil rights, and internationalism. The split may occur along class lines:

> The fundamental factor in noneconomic liberalism is not actually class, but education, general sophistication, and probably to a certain extent psychic security. But since these factors are strongly correlated with class, noneconomic liberalism is positively associated with social status (the wealthier are more tolerant), while eco-

> nomic liberalism is inversely correlated with social status (the poor are more leftist on such issues).[65]

But often members of the same class may be pulled in opposite directions. When Thurman Arnold was maintaining that "freedom of opportunity is the great value that must be preserved above all others," Chester Bowles was arguing that "there can be no real freedom in the presence of economic insecurity."[66] The tension between the two kinds of liberalism may exist in a single individual, as is shown by Reinhold Niebuhr's remarks on the Soviet Union shortly after World War II:

> Western civilization has not solved its economic problem within the framework of a free society; and it may not be able to. It may be that Russia has solved the economic problem. We ought at any rate to appreciate the real achievements of this great adventure. But if genuine liberty has been sacrificed for the sake of achieving its end, we ought to know about it.[67]

It has been argued that in striving for both individual freedom and economic security "the omnibus of liberalism has been heading toward two objectives which no other way of thought and action has even tried to reach simultaneously, and which most systems consider ultimately irreconcilable."[68] The tension between these two goals of liberalism is reflected in the varying attitudes among liberals toward Communism. Economic liberals are likely to be so impressed by the alleged economic equality in the Soviet Union that they overlook the lack of intellectual freedom there.[69] They may agree with Robert Heilbroner that "revolution, authoritarianism, and collectivism are often the *only* instruments by which essential social changes can be made."[70] The noneconomic liberal, who is concerned about civil liberties both at home and abroad, is more likely to be explicitly anticommunist.

Thus it seems as if liberalism, already besieged from the right and the left, is also divided within itself between the advocates of bread and the advocates of freedom. Since exaggeration is endemic in political rhetoric, both kinds of liberals tend to advocate their goals in absolute terms, and neither is fully satisfied with compromises that aim for the maximum of both freedom and security without too greatly jeopardizing either. It is small wonder, then, that liberals are continually pondering aloud, "Just exactly what is true liberalism, anyway?"

CRITICISMS OF LIBERALISM FROM OUTSIDERS

The liberal is always analyzing himself—intellectuals enjoy being introspective, and a meaty article on "The Plight of the Liberal Today" can always be sandwiched in among the girdle ads in the *New York Times Magazine*—but the analysis is not complete without a look at some criticisms of liberalism from various outsiders.

One outsider is the black, who for several years has been openly critical of the white liberal.[71] Although for most of the postwar period liberals have fought for civil rights, liberalism must be regarded as a Johnny-come-lately to the fight for racial justice. The Populist and Progressive record on racial and ethnic questions is nothing for liberals to brag about. The executive order of 1941 forbidding discrimination in war industries was signed by President Roosevelt only after strong pressure from Philip Randolph. The ten-point creed for liberals published by ADA chairman Wilson Wyatt in 1947 made no mention of civil rights. When civil rights became a liberal cause in the 1950s, some liberals tried to compensate for their previous neglect by switching to a fervid advocacy that many blacks liked no better than the neglect.[72]

In 1964 *Commentary* sponsored a round-table discussion, "Liberalism and the Negro." The round table was composed of one black and three whites—and the latter did most of the talking. Kenneth Clark, a member of the audience, commented as a self-proclaimed "former liberal": "I must confess that I now see white American liberalism primarily in terms of the adjective 'white.' "[73] James Baldwin complained, "The liberal assumption is that once you arrive at a certain level of social and economic status in American life, there's nothing left to worry about."[74]

A number of the younger black militants at the end of the sixties were rejecting Western liberalism and its "rationality" in favor of a nonverbal black culture that was based on experience and was aware of irrationality.[75] Their position in many ways is akin to the rejection by the nineteenth-century Romantics of the superficial rationality of the Enlightenment—what Blake in "Los the Terrible" called "single vision and Newton's sleep" and Emerson in *Nature* dismissed as mere "understanding." Like such radical critics as Theodore Roszak, they reject a central tenet of liberalism—the belief in the sufficiency of scientific "reason" to solve all human problems.

A second outsider is the non-Westerner. Both Tolstoy and Dostoevsky, who represent opposing trends in the nineteenth-century Russian novel, agree in their criticism of Western liberalism. Tolstoy satirizes at some length the "liberal" Stepan Arkadyevitch Oblonsky in the early pages of *Anna Karenina*, just as Dostoevsky makes fun of the shallow-minded liberal, Miüsov, in the opening chapters of *The Brothers Karamazov.* Dostoevsky then proceeds to examine critically three liberals: Rakitin, the seminary student turned journalist, who is the epitome of the liberal young man on the make; Ivan, the intellectual brother whose "Euclidian mind" has been powerfully shaped by Western rationalism; and Ivan's fictional creation, the Grand Inquisitor, a portrayal of the liberal intellectual corrupted by possession of absolute power, who, in the name of "universal human happiness," has abolished what he sees as the one cause of unhappiness—human freedom.

The Grand Inquisitor represents the reductio ad absurdum of the views of the "economic" liberal, and Dostoevsky can be accused of stacking the cards against him. Nevertheless, insofar as the liberal is tempted to play God in order to make men happy, he runs the risk of resembling, to some degree, this creature of Dostoevsky's fantasy. The resemblance is plain in W. H. Auden's comic equivalent of the Inquisitor, the character of Herod in *For the Time Being*, who exclaims pathetically that he is a liberal who only wants to make everybody happy.[76]

Another outsider is the conservative. Probably the best-known—and certainly the wittiest and most prolific—conservative critic of liberalism is William F. Buckley, editor of the *National Review*.[77] A more scholarly brand of criticism of liberalism can be found in the writings of Peter Viereck and Clinton Rossiter, academic conservatives who disagreed with the *National Review*'s kind of conservatism and were consequently sometimes accused of really being liberals in disguise.[78] Both men speak as friends of liberalism, rather than as enemies, and both stress that liberals and conservatives have more in common with each other than with extremists at either end of the spectrum. Rossiter finds the differences between conservatives and liberals to consist chiefly in their temperament, their views of human nature, and the relative emphasis they place on freedom and order; he quotes with approval Emerson's remarks on liberalism and conservatism: "each is a good half, but an impossible whole. . . . In a true society, in a true man, both must combine."[79]

The approach taken by Rossiter and Viereck runs the risk of blurring the differences between liberalism and conservatism—which both men insist are real ones. But there is also the paradox stated by Adlai Stevenson in 1952 (during a campaign in which the Republicans were saying, "It's time for a change," and the Democrats were countering, "You never had it so good"): "The strange alchemy of time has somehow converted the Democrats into the truly conservative party of this country—the party dedicated to conserving all that is best, and building solidly and safely on these foundations."[80] Even if politically motivated, Stevenson's comment contained a core of truth. Liberals in the early 1950s were indeed becoming more conservative, partly because of prolonged prosperity, partly because of the obvious Communist threat, and partly because they felt that a good measure of reform was being achieved.[81] And thus, as liberals and conservatives moved toward a loose consensus in the 1950s and early 1960s, the younger and more passionate advocates of social reform felt they had nowhere to go but to radicalism.

Criticism of liberalism from the radical, a fourth outsider, is far more dangerous to the liberal than is criticism from the conservative. The liberal finds the conservative a comfortable enemy; he finds the radical an uncomfortable friend. The conservative's impulses carry him in the opposite direction from the liberal's. The conservative is for standing pat; the liberal is for change. The conservative wishes to preserve the blessings of freedom; the liberal wants to expand them. The conservative looks at man's weaknesses; the liberal looks at his strengths. The conservative looks to the past; the liberal looks to the future. Although liberals and conservatives can sometimes ally themselves against a common enemy, the liberal does not feel drawn either emotionally or ideologically to the position of his favorite adversary.

The radical's impulses go in the same direction as the liberal's, and they are stronger. Hence he has for the liberal somewhat the attraction that the street-corner tough has for the timid boy down the block. Eric Sevareid, in describing his reactions upon meeting a young Communist activist in Chungking during World War II, has expressed this feeling well:

> Again I had the old feeling of uselessness, the crushing sense of ineffectualness that comes to the vacillating liberal, who contests

> only with words, in the presence of a truly strong, dedicated person who has accepted the perils of action, made his decision and cast his personal life into the account as a thing of value only in terms of the future he himself will never see.[82]

If the liberal is attracted by the radical's toughness and commitment, the radical has little use for what he regards as the liberal's moral shilly-shallying. "A liberal," Heywood Broun has been quoted as saying, "is a man who leaves the room when the fight begins."[83] There is something about the liberal's tentativeness, his balanced assessment of situations, and his willingness to compromise that drives the radical into a fury. Where the liberal prides himself on being pragmatic, the radical prides himself on being doctrinaire, and he frequently charges that the liberal's preoccupation with means may lead him to a neglect of ends.[84]

The liberal and the radical share a common concern for social justice, but the radical can complain that the liberal's concern is a shallow one, because the radical, by definition, goes to the root of injustice.[85] Not all radicals agree, however, on where that root is to be found. Many of them find the root of social ills to lie in institutions. Like Rousseau, they see man as good by nature, but corrupted by society. And their answer is simple: destroy the evil institutions, and man's natural goodness will reassert itself. William Lloyd Garrison, the militant nineteenth-century abolitionist, is a good representative of this radical approach. Despite his deeply held pacifist beliefs, his rhetoric is one of violence and destruction. Addressing a personified "Oppression," he declaims in the first issue of his newspaper *The Liberator*:

> I swear, while life-blood warms my throbbing veins,
> Still to oppose and thwart, with heart and hand,
> Thy brutalizing sway—till Afric's chains
> Are burst, and Freedom rules the rescued land,—
> Trampling Oppression and his iron rod:
> Such is the vow I take—SO HELP ME GOD![86]

Because his thinking and his rage were so concentrated on the destruction of slavery as an institution, he spared relatively little thought for the matter of how the freed slave was to be given the training, education, and other practical help he would need to live successfully in freedom.

Garrison represents one kind of radicalism—what we might call "political radicalism." But another radical approach, which might be called "moral radicalism," is exemplified by one of Garrison's contemporaries, Henry David Thoreau, and by an earlier opponent of slavery, John Woolman. This kind of radicalism sees the root of social evils as located not in institutions but in the human heart. As Thoreau put it:

> There are a thousand hacking at the branches of evil to one who is striking at the root, and it may be that he who bestows the largest amount of time and money on the needy is doing the most by his mode of life to produce that misery which he strives in vain to relieve. It is the pious slaveholder devoting the proceeds of every tenth slave to buy a Sunday's liberty for the rest.[87]

Woolman, one of the most persistent and effective opponents of slavery in eighteenth-century America, continually identified the source of slavery —and of racial prejudice, which was not eradicated by the abolition of slavery—as selfishness in the human heart and mind:

> Selfishness being indulged, clouds the understanding; and where selfish men, for a long time, proceed on their way without opposition, the deceitfulness of unrighteousness gets so rooted in their intellects, that a candid examination of things relating to self-interest is prevented; and in this circumstance some who would not agree to make a slave of a person whose colour is like their own, appear easy in making slaves of others of a different colour, though their understandings and morals are equal to the generality of men of their own colour.[88]

Just as the source of evil lay in the human heart and mind, so, according to Woolman, did the remedy for that evil:

> There is a principle which is pure, placed in the human mind, which in different places and ages hath had different names. It is, however, pure and proceeds from God. It is deep and inward, confined to no forms of religion nor excluded from any, where the heart stands in perfect sincerity. In whomsoever this takes root and grows, of what nation soever, they become brethren in the best sense of the expression.[89]

Because political radicals and moral radicals often share the same urgent and uncompromising concern for social justice, the difference between them is not often taken into account.[90] But the difference can be real, as Hawthorne shows in his short story, "Earth's Holocaust." He pictures a vast bonfire in which all of the allegedly corrupt human institutions are incinerated by a crowd of ardent reformers. At the end, one of the spectators of the bonfire comments grimly to the narrator that nothing in the world will change for the better, because the human heart remains as the source of all evil. The narrator muses:

> How sad a truth, if true it were, that man's agelong endeavor for perfection has served only to render him the mockery of the evil principle, from the fatal circumstance of an error at the very root of the matter! The heart, the heart,—there was the little yet boundless sphere wherein existed the original wrong of which the crime and misery of this outward world were merely types.

Where the political radical in his hatred of institutions runs the risk of falling in love with the act of destruction itself—like the mad Professor in Conrad's *The Secret Agent*, with the bomb continually in his pocket—the moral radical is more likely to say, as Whitman does in "I Hear It Was Charged Against Me":

> I hear it was charged against me that I sought to
> destroy institutions.
> But really I am neither for nor against institutions,
> (What indeed have I in common with them? or what with
> the destruction of them?)

The factor that enables the moral radical to maintain his perspective on institutions often seems to be an overwhelming sense of a God who is infinitely larger, more powerful, more permanent, and ultimately more real than any human institutions. Woolman, a Quaker, uses fairly orthodox religious—though not sectarian—language. Thoreau, who "signed out" of church, disliked clergymen, and shocked the pious by his religious unorthodoxy, nevertheless bases his doctrine of civil disobedience on the possibility of appealing from men to their Maker. Toward the end of "Civil Disobedience," he speaks of an ultimate source of truth beyond the commonly accepted institutions:

> They who know of no purer sources of truth, who have traced up its stream no higher, stand, and wisely stand, by the Bible and the Constitution, and drink at it there with reverence and humility; but they who behold where it comes trickling into this lake or that pool, gird up their loins once more, and continue their pilgrimage toward its fountain-head.

Lincoln, who appears to have shared Thoreau's lack of orthodoxy, had a view of God similar to that of Thoreau and Woolman. In a private memorandum probably written after battlefield reverses in September 1862, Lincoln meditated:

> The will of God prevails. In great contests each party claims to act in accordance with the will of God. Both *may* be, and one *must* be wrong. God can not be *for* and *against* the same thing at the same time. In the present civil war it is quite possible that God's purpose is something quite different from the purpose of either party—and yet the human instrumentalities, working just as they do, are of the best adaptation to effect His purpose.[91]

Lincoln's best-known expression of this idea is found in his Second Inaugural Address, which has been described as "almost a perfect model of the . . . task of remaining loyal and responsible toward the moral treasures of a free civilization on the one hand while yet having some religious vantage point over the struggle."[92]

The outsiders who criticize liberalism—blacks, non-Westerners, conservatives, and radicals—all feel deeply that it is lacking something; they cannot always articulate what that something is. Dwight Macdonald, who admitted in the 1950s, "I just don't seem to have the knack for religious experience," nevertheless went on to speculate "from a purely intellectual point of view" that the missing factor might be God, for

> I have lost confidence in the dominant non-religious social tendency in the country today: the Marx-cum-Dewey approach represented by Sidney Hook (pure), the liberal weeklies (debased), the Reuther brothers and Senator Humphrey ("grass roots"), the Americans for Democratic Action (official) and *Partisan Review* (highbrow). This seems to me to have failed politically, culturally, and even scientifically.[93]

It is hard to discount Macdonald's hypothesis. Lacking a personal, transcendent, and eternal source for values, for historical perspective, and for personal humility, it seems all too easy for liberals to succumb to what Senator Fulbright described in the 1960s as "the arrogance of power."

Liberalism's tragic flaw, then, may prove to be its insistence on being secular, rationalistic, and humanistic. As an attempt to achieve by secular means such religious ends as peace, brotherhood, and economic and social justice, liberalism's reaction to the social conservatism so often shown by religious institutions is easily understandable.[94] But in abandoning these institutions, liberalism has been all too prone to abandon its own religious roots as well. Without these roots, liberalism is vulnerable. When science and technology seem to be leading mankind into a dismal future, the liberal has no ready alternative, for he has pinned his faith on science and technology as an answer to human problems. When reason seems too feeble an instrument to cope with the gigantic irrationalities of modern society, the liberal is baffled, for he has pinned his hopes on reason. When an idealistic, well-intentioned, pragmatic foreign policy leads into the morass of Vietnam, the liberal is confused, for idealism, good intentions, and pragmatism have long been his guideposts.

Liberalism's problem is not that it is a bad faith; its problem may be rather that it is a good faith—almost the best of faiths. But to reverse Voltaire's maxim, the good may be the enemy of the best. To meet the new challenges it faces, it may not be enough for liberalism to reiterate its venerable dogmas, no matter how valuable they have been in the past. The approach that I have called moral radicalism suggests a standard by which liberalism can be judged and also a direction it could take in the future. Whether it chooses to take this direction depends, of course, not on the abstraction called liberalism but on the decisions of the living people who call themselves liberals.

THE LIBERAL COMMUNITY

Just who are these liberals? The liberal does not exist in isolation: he is conscious of being part of both a tradition and a living community. This community cannot be confined merely to the membership of an official liberal organization like the ADA, nor to the loose geographical and cultural cluster that has been dubbed "the New York intellectual community."[95] It is both larger and less sharply defined. In the late 1960s the term "liberal establishment," which had the advantage of being both vague and pejorative, gained increasing vogue among critics of liberalism from both the right and the left.[96] Richard Rovere, who set out to parody the conservative notion of a liberal establishment, seemed to end by being a bit attracted to the idea.[97] John Kenneth Galbraith, in an attack on the Vietnam War in 1966, spoke of an "establishment" that only a few years before would have seemed the product of conservative fantasies, saying that

> it was also agreed that foreign policy would be entrusted to the permanent diplomatic and military establishment under the general chairmanship of the New York foreign policy syndics. The latter—the Dulles-McCloy-Lovett communion with which, I am sure, Dean Rusk would wish to be associated and of which Dean Acheson is a latter day associate—has now provided the grace notes for American foreign policy for twenty years.[98]

The popularity of such terms as "Establishment," "power elite," "power structure," and "military-industrial complex" —vague as they are, and loosely used as they are—indicates a persistent feeling that there is at least a common set of beliefs or interests that links many of the leading figures in politics, business, labor, education, the military, and the news media. There seems to be enough of a common sense of purpose and common interpretation of the world among the leaders of these groups to constitute a community. And a majority of the members of this community would probably call themselves "liberals"—with the appropriate qualifying adjectives.

Within this liberal community are perhaps two hundred prestigious opinionmakers, most of whom are closely tied to the leading intellectual journals, and half of whom live within fifty miles of New York City.[99] But the liberal community cannot be restricted to the Eastern seaboard or to any such elite. Its ordinary rank-and-file members

> can be found everywhere, and very good people they are—civic-minded, hard-working in thankless causes, the people you can count on. They take the popular side on local issues: work against the real estate lobby to get public housing, fight racial discrimination, join committees to expose vice and corruption. Like most liberals everywhere, they tend to be a little

> self-righteous about their independence and intelligence, but they can be forgiven their foibles.[100]

The members of this community derive their common interpretation of the world from the news media that provide them with their information and ideas, and most perceptive journalists are aware of this fact. When James Reston was complimented for his ability to change the thinking of government leaders, he replied that he did not write for them at all; he wrote "for the lonely professor at some college in a small town in the Midwest, who cared and wanted some of the complicated issues of the day clarified by a friend."[101]

Members of the liberal community are most likely to derive their sense of belonging to a community from the magazines they read. Magazines are edited for "little publics within the population as a whole," and they come to be identified with some particular point of view.[102] The spectrum of magazines read by members of the liberal community at the end of the 1960s may be indicated by a representative group of ten—five with circulations under 150,000 and five with circulations in excess of 300,000.[103] The first group includes the two that are always named in any discussion of liberal magazines, the venerable *Nation* and *New Republic*—founded in 1865 and 1914 respectively—even though Robert Sherrill, who had written for the *Nation* since 1964, claimed in 1969 that the magazine would no longer designate itself "liberal."[104] Between 1945 and 1969 the *Nation*'s circulation declined from 42,000 to about 30,000. By contrast, the *New Republic* soared from a circulation of 38,000 in 1945 to a spectacular 145,000 by 1969. The increase was not continuous: under the editorship of Henry Wallace it had jumped to 87,000 in 1948 and then dropped to a nadir of 24,000 in 1952. The advent of the New Frontier began a second period of rapid circulation growth: from 36,000 in 1960 to 72,000 in 1962 and 120,000 in 1966. This first group might also include two small but persistent political magazines—the *Progressive* and the *New Leader*—with circulations that were close to that of the *Nation*, and a cultural magazine, *Commentary*, with a circulation that had grown to more than 63,000 by the end of the 1960s.[105]

The second group of magazines read by members of the liberal community would not necessarily all be given the label of "liberal" by every student of journalism—partly because they are primarily cultural rather than political in their orientation and partly because they are commercially successful. (There is a persistent tendency to regard liberal journalism as synonymous with unprofitable journalism—a frequent enough occurrence, unfortunately, to make such identification plausible.)[106] The magazines in this second group, which would include the *New Yorker, Atlantic, Harper's, Saturday Review*, and *Time*, all grew dramatically in circulation between 1945 and 1969. The *New Yorker* and *Atlantic* each more than doubled their circulation—to 482,000 and 325,000 respectively. *Harpers* and *Time* more than tripled their circulations—from 114,000 to 381,000 and from 1.2 million to 4.2 million respectively. *Saturday Review*, which broadened its coverage to include education, science, and communications, increased its circulation nearly twelve times—from 50,000 to over 590,000 in 1969.[107] *Time*, whose founder once designated it—rather defensively—as "liberal" and whose parent organization has employed an impressive roster of liberals over the years, found common ground with liberals on such issues as opposition to Senator Joe McCarthy and support for world law, the United Nations, and the more respectable elements of the civil-rights movement.[108] And though it was long fashionable for liberals to curse *Time*, they nevertheless seemed to read it.

In this spectrum of magazines, *The Reporter*, a self-proclaimed "liberal magazine" that appeared for nineteen years, from 1949 to 1968, would stand somewhere in the middle. Neither a weekly nor a monthly, it was unusual in being published every two weeks. A "quality" magazine like those of the second group, it was nevertheless financially unsuccessful like many of the first group. Its peak circulation of over 200,000 at the time of its demise was greater than the then circulation of the *Nation, New Republic,* and *Progressive* combined; but it was substantially less than that of any of those in the second group. *The Reporter*'s staff included a former managing editor of the *Nation*, Robert Bendiner, and two former writers for the *New Republic*, William Harlan Hale and Claire Sterling. Its contributors included staff members of *Time,* the *New Yorker,* and the *Progressive*; several of its contributors also contributed regularly to *Commentary*. When *The Reporter* ceased publication in 1968, several of its staff went briefly to *Harper's,* which also took over its unexpired subscriptions.

The magazine, it has been claimed, is the most personal form of journalism: "one man can influence every idea, every layout, every word that appears in print."[109] The history of *The Reporter,* with its brilliant staffers and many well-known contributors, tends to confirm this observation; it can justifiably be seen, in many ways, as the lengthened shadow of a single man. An examination of *The Reporter* must begin with a look at the ideas of its founder, editor, and publisher—Dr. Max Ascoli.

NOTES

1. John Dewey, *Liberalism and Social Action* (New York: Putnam's, 1935), p. 1.

2. Alfred Kazin, *Starting Out in the Thirties* (Boston and Toronto: Little, Brown, 1965), p. 87. Cf. Granville Hicks, *Where We Came Out* (New York: Viking, 1954); James Wechsler, *The Age of Suspicion* (New York: Random House, 1953); Dwight Macdonald, *Memoirs of a Revolutionist* (New York: Farrar, Straus, and Cudahy, 1957); Daniel Aaron, *Writers on the Left* (New York: Harcourt, Brace and World, 1961); and Frank A. Warren III, *Liberals and Communism* (Bloomington and London: Indiana University Press, 1966).

3. Corey Ford, *Donovan of OSS* (Boston and Toronto: Little, Brown, 1970), pp. 135, 164, 320.

4. Macdonald, pp. 292-96.

5. Clifton Brock, *Americans for Democratic Action: Its Role in National Politics* (Washington, D.C.: Public Affairs Press, 1962), pp. 49-52. Wechsler, pp. 211-17.

6. See Palmer W. Wright, "The 'New Liberalism' of the Fifties: Reinhold Niebuhr, David Riesman, Lionel Trilling and the American Intellectual" (Ph.D. dissertation, University of Michigan, 1966).

7. Lionel Trilling, *The Liberal Imagination* (New York: Viking, 1950), p. vii.

8. Brock, p. 118.

9. Ibid., p. 15. Cf. James MacGregor Burns, *John Kennedy: A Political Profile* (New York: Harcourt, Brace, 1960), pp. 132-36.

10. Arthur M. Schlesinger, Jr., *Kennedy or Nixon: Does It Make Any Difference?* (New York: Macmillan, 1960), p. 26.

11. Andrew Knight, "America's Frozen Liberals," *Progressive,* February 1969, p. 33.

12. Brock, pp. 12, 174.

13. See Eric F. Goldman, *The Tragedy of Lyndon Johnson* (New York: Knopf, 1969).

14. Arthur M. Schlesinger, Jr., *The Vital Center* (Boston: Houghton Mifflin, 1949).

15. W. H. Auden, "A Russian Aesthete," *New Yorker,* April 4, 1970, p. 136.

16. Harold Laski, *The Rise of Liberalism* (New York and London: Harper and Brothers, 1936), p. 5. A useful recent discussion of liberalism, with a lengthy bibliography, is William Gerber's *American Liberalism* (Boston: Twayne, 1975).

17. Trilling, pp. vii-viii.

18. David G. Smith, "Liberalism," *International Encyclopedia of the Social Sciences* (New York: Macmillan and the Free Press, 1968), vol. 9, p. 276.

19. Brock, p. 23.

20. Quoted in Eric F. Goldman and Mary Paull, "Liberals on Liberalism," *New Republic,* July 22, 1946, p. 71.

21. Adlai Stevenson, "The Mission of Liberalism," *New Republic,* September 24, 1956, p. 11.

22. Joseph C. Harsch, "Are Liberals Obsolete?" *The Reporter,* September 30, 1952, p. 15.

23. Ibid., p. 16.

24. Eric F. Goldman, "The American Liberal: After the Fair Deal, What?" *The Reporter,* June 23, 1953, p. 25.

25. Amitai Etzioni, "Neo-Liberalism—the Turn of the '60's," *Commentary,* December 1960, pp. 473-79.

26. Hicks, p. 246. Sidney Hook, et al., "Liberal Anti-Communism Revisited: A Symposium," *Commentary,* September 1967, p. 45. Robert Bendiner, "What Kind of Liberal Are You?" *Commentary,* September 1949, pp. 238-42. Arnold Kaufman, *The Radical Liberal* (New York: Atherton Press, 1968). Bendiner lists a fascinating variety of liberals, including such subspecies as the Southern Liberal, the Hyperthyroid Liberal ("often useful, but hard to live with"), and the Western Maverick.

27. *Time,* February 21, 1949, p. 24.

28. Bendiner, p. 238. Emphasis in the original.

29. Kaufman, p. 15. Dewey, p. 62.

30. Wechsler, pp. 216-17.

31. David Lilienthal, *Journals,* vol. 3, *The Venturesome Years, 1950-1955* (New York: Harper and Row, 1966), pp. 384-85.

32. Robert Bendiner, "What Is a Liberal?" *Nation,* March 26, 1949, pp. 349-50.

33. See John Brooks, *Business Adventures* (New York: Weybright and Talley, 1969), pp. 249-75.

34. Harsch, p. 16.

35. See Daniel Bell, *The End of Ideology* (New York: The Free Press, 1961), pp. 299-314, 393-407.

36. Michael Harrington, "Liberalism According to Galbraith," *Commentary,* October 1967, p. 77.

37. See, e.g., Theodore J. Lowi, *The End of Liberalism* (New York: Norton, 1969), a book proposing the replacement of "interest-group liberalism" with "juridical democracy."

38. Schlesinger, *The Vital Center,* pp. 143-46. Cf. Eric Hoffer, *The True Believer* (New York: Harper, 1951), pp. 72-74, 84-85; and Clinton Rossiter, *Conservatism in America* (New York: Knopf, 1962), pp. 10-15.

39. Quoted in Bendiner, "What Is a Liberal?" *Nation,* March 26, 1949, pp. 349-50.

40. Quoted in Goldman and Paull, p. 71.

41. Brock, p. 52.

42. Wilson Wyatt, "Creed for Liberals: A Ten-Point Program," *New York Times Magazine,* July 27, 1947, p. 36. Eric Goldman, *The Crucial Decade—and After* (New York: Vintage, 1960), pp. 158-60.

43. "Six Liberals Define Liberalism," *New York Times Magazine,* April 19, 1959, p. 82.

44. John Kenneth Galbraith, "An Agenda for American Liberals," *Commentary,* June 1966, pp. 29, 31.

45. "Liberal Anti-Communism Revisited: A Symposium," *Commentary,* September 1967, p. 52.

46. Ibid., p. 67.

47. Ibid., p. 54.

48. Ibid., p. 45.

49. See Townsend Hoopes, *The Limits of Intervention* (New York: David McKay, 1969), p. 240; Kenneth O'Donnell, "LBJ and the Kennedys," *Life,* August 7, 1970, pp. 51-52; Theodore Sorenson, *The Kennedy Legacy* (New York: Macmillan, 1969), pp. 202-17. For a more hawkish interpretation of Kennedy's Vietnam policy see Sorenson, *Kennedy* (New York: Harper and Row, 1965), pp. 660-61.

50. See, e.g., Carl Oglesby, "Bankruptcy of the Liberals," *Commonweal,* January 7, 1966, pp. 396-400.

51. Robert Osgood, *Ideals and Self-Interest in America's Foreign Relations* (Chicago: University of Chicago Press, 1953), pp. 300-80.

52. Eric Goldman, "The Liberals, the Blacks, and the War," *New York Times Magazine,* November 30, 1969, p. 53.

53. Robert Paul Wolff, *The Poverty of Liberalism* (Boston: Beacon Press, 1968).

54. William James, *Pragmatism* (London: Longmans Green, 1907), p. 6.

55. Chester Bowles, as quoted in Goldman and Paull, p. 71.

56. Goldman, "The Liberals, the Blacks, and the War," p. 40. Cf. Willson H. Coates and Hayden V. White, *The Ordeal of Liberal Humanism* (New York: McGraw-Hill, 1970), p. 447.

57. Merrill Peterson, *The Jefferson Image in the American Mind* (New York: Oxford University Press, 1960). Cf. Leonard W. Levy, ed., *Freedom of the Press from Zenger to Jefferson* (Indianapolis: Bobbs-Merrill, 1966), pp. 327-76; and Clinton Rossiter, "Which Jefferson Do You Quote?" *The Reporter,* December 15, 1955, pp. 33-36.

58. Charles Forcey, *The Crossroads of Liberalism* (New York: Oxford University Press, 1961), p. 29.

59. Schlesinger, *The Vital Center,* p. 16.

60. Frederick C. Howe, *The Confessions of a Reformer* (New York: Charles Scribner's Sons, 1925), p. 322.

61. Charles Frankel, "A Liberal Is a Liberal Is a ——," *New York Times Magazine,* February 28, 1960, p. 82.

62. "Liberal Anti-Communism Revisited," pp. 44-45.

63. Seymour M. Lipset, *Political Man* (New York: Doubleday, 1960). pp. 314-18.

64. Theodore Roszak, *The Making of a Counter Culture* (Garden City, N.Y.: Doubleday, 1969), especially pp. 42-55, 205-38.

65. Lipset, pp. 101-02, 298.

66. Both quoted in Goldman and Paull, pp. 71-72.

67. Reinhold Niebuhr, "The Russian Adventure," *Nation,* February 23, 1946, p. 234.

68. Goldman and Paull, p. 72.

69. See, e.g., the example cited by Joseph Wood Krutch in his *More Lives Than One* (New York: William Sloane Associates, 1962), p. 247.

70. Robert L. Heilbroner, "Making a Rational Foreign Policy Now," *Harper's,* September 1968, p. 69. Emphasis in the original.

71. See, e.g., Loren Miller, "Farewell to Liberals: a Negro View," *Nation,* October 20, 1962, pp. 235-38. The "white liberal" is not a new phenomenon: his fictional prototype—well-intentioned, naive, moralistic, condescending, and ultimately somewhat ridiculous—may be seen in Harriet Beecher Stowe's satirical portrait of Miss Ophelia in *Uncle Tom's Cabin.*

72. Goldman, "The Liberals, the Blacks and the War," pp. 48, 50; Wyatt, "Creed for Liberals: A Ten-Point Program," p. 36; Robert Penn Warren, *Who Speaks for the Negro?* (New York: Vintage, 1966), pp. 436-40.

73. James Baldwin, Nathan Glazer, Sidney Hook, and Gunnar Myrdal, "Liberalism and the Negro: A Round Table Discussion," *Commentary,* March 1964, p. 39.

74. Ibid., p. 38.

75. See, e.g., Julius Lester, *Look Out, Whitey! Black Power's Gon' Get Your Mama!* (New York: Dial Press, 1968).

76. *The Collected Poetry of W. H. Auden* (New York: Random House, 1945), p. 460.

77. See, e.g., his *Up from Liberalism* (New York: McDowell, Obolensky, 1959). Buckley was one of the sixty-four writers who contributed to the symposium, "What Is a Liberal—Who Is a Conservative?" *Commentary,* September 1976, pp. 31-113.

78. Peter Viereck, *Conservatism Revisited,* rev. ed. (New York: The Free Press, 1962), p. 141; Rossiter, *Conservatism in America,* pp. vii-viii.

79. Rossiter, *Conservatism in America,* pp. 56-57.

80. Quoted in Viereck, p. 156.

81. Lipset, pp. 341-42.

82. Eric Sevareid, *Not So Wild a Dream* (New York: Knopf, 1946), p. 329.

83. George Seldes, comp., *The Great Quotations* (New York: Lyle Stuart, 1960), p. 116.

84. See Noam Chomsky, *American Power and the New Mandarins* (New York: Pantheon, 1969), p. 5.

85. See, for example, the comparison made by Sinclair Lewis in *Main Street* (Chapter 21) between his heroine Carol Kennicott, a radical, and her friend Vida Sherwin, a liberal. Cf. Gerber, pp. 82-83.

86. Quoted in John Jay Chapman, *William Lloyd Garrison* (New York: Moffat, Yard, 1913), pp. 40-41.

87. *Walden,* Chapter 1, "Economy." Thoreau's judgment of contemporary reformers could be scathing. See his *Journal* entry for June 17, 1853. Emerson's "Self-Reliance" and his "Ode: Inscribed to W. H. Channing" express similar criticism of reformers.

88. John Woolman, *The Journal and Major Essays of John Woolman*, ed. Phillips P. Moulton (New York: Oxford University Press, 1971), p. 225.

89. Ibid., p. 236.

90. Staughton Lynd's *Intellectual Origins of American Radicalism* (New York:

Random House, 1969), for instance, makes no attempt to distinguish between the approach of Garrison and that of Thoreau and Woolman.

91. Quoted in William J. Wolf, *The Almost Chosen People* (Garden City, N.Y.: Doubleday, 1959), p. 147.

92. Reinhold Niebuhr, *The Irony of American History* (New York: Scribner's, 1952), p. 172.

93. Macdonald, pp. 370-71.

94. See Alexis de Tocqueville, *Democracy in America,* ed. Phillips Bradley (New York: Vintage, 1958), vol. 1, pp. 12-13.

95. Christopher Lasch, *The New Radicalism in America* (New York: Knopf, 1965), pp. 319-22.

96. See e.g., M. Stanton Evans, *The Liberal Establishment* (New York: Devin-Adair, 1965), an earnest, dogged indictment by a conservative. David Halberstam's *The Best and the Brightest* (New York: Random House, 1972) looks at many of the same liberals from a different—but equally critical—perspective. For a discussion of the origins of the modern use of "establishment," see Henry Fairlie, "Evolution of a Term," *New Yorker,* October 19, 1968, pp. 173-206.

97. Richard Rovere, *The American Establishment and Other Reports, Opinions, and Speculations* (New York: Harcourt, Brace and World, 1962), pp. 3-21. See the comment on Rovere in Lasch, p. 322.

98. Galbraith, "An Agenda for American Liberals," p. 31.

99. See Charles Kadushin, *The American Intellectual Elite* (Boston: Little, Brown 1974).

100. Hicks, p. 173.

101. David Halberstam, "Love, Life, and Selling Out in Poland," *Harper's,* July 1967, p. 80.

102. Theodore Peterson, *Magazines in the Twentieth Century,* 2nd ed. (Urbana, Ill.: University of Illinois Press, 1964), pp. 13-14.

103. Circulation figures are taken from various annual editions of Ayer's *Directory of Newspapers and Periodicals* and are rounded to the nearest thousand.

104. Robert Sherrill, "Weeklies and Weaklies," *Antioch Review,* Spring 1969, p. 28.

105. For an account by its editor of the creation of the "new" *Commentary* in the early 1960s, see Norman Podhoretz, *Making It* (New York: Random House, 1967).

106. See George Soule, "Liberal Journalism: A Diagnosis," *Yale Review,* Winter 1950, pp. 326-36.

107. For the ironic aftermath of this "success" see William H. Honan, "The Morning After the Saturday Review," *Esquire,* November 1973, pp. 176-81, 208, 212, 214, 216; and Dwight Macdonald, "Norman Cousins' Flat *World*," *Columbia Forum,* Fall 1972, reprinted in his *Discriminations* (New York: Grossman, 1974), pp. 174-93.

108. Robert T. Elson, *Time Inc.* (New York: Atheneum, 1968), pp. 358-59. Kazin, pp. 105, 111. Joseph Epstein, "Henry Luce and His Time," *Commentary,* November 1967, pp. 41-42.

109. Clay S. Felker, "Life Cycles in the Age of Magazines," *Antioch Review,* Spring 1969, p. 7.

Chapter Two

The Liberalism of Max Ascoli

In all my writings and teaching I am afraid there has been a consistency which shows the hopeless limitations of my mind.
Max Ascoli, 1952

It is impossible to discuss the ideas of Max Ascoli without also discussing his personality—a powerful one—and impossible to discuss either without considering his background. "You must remember," says one of his friends, "that Max is Italian."[1] For anyone who has heard him speak, that fact is impossible to forget. More than forty years of living in the United States have not erased his thick and sometimes impenetrable accent—an accent that some of his staff referred to as "Ascolese." ("I wonder why I still have this accent?" he mused aloud early in 1971.)[2] He had for years both an English-speaking and an Italian-speaking secretary working together in his outer office, and one former member of *The Reporter*'s staff recalls Ascoli uttering such Goldwynisms as, "I think this will sell like wildcakes," or "I want you to go through this copy with a fine-tooth brush."

Yet if he has remained indelibly Italian, he is at the same time fiercely proud of being a "self-made American," having become a United States citizen in 1939.[3] "I am more American than you are," he once told an American member of his staff who had lived for many years in Europe.

He is a Jew who in some respects may be described as profoundly Catholic, a liberal who has found common ground on various issues over the years with both conservatives and radicals, a serious and somewhat ponderous political writer who in private conversation can be charming, witty, and unexpectedly droll. Max Ascoli, in short, appears to be a bundle of paradoxes.

"As well as I think I know him," says one of his friends, "Ascoli remains a mystery to me. He's a terribly complex person." But some of these complexities may fall into perspective if he is compared with that paradoxical eighteenth-century English personality, Dr. Samuel Johnson—a comparison Ascoli probably would not like, because an Italianate or American Dr. Johnson may be difficult to imagine. But if one could be imagined, he would share a number of characteristics with Ascoli. He would be a man in continual tension between a powerful intellect and passionate feelings, a man possessed of deep and genuine religious convictions but also capable of great rudeness and arrogance toward his associates, a man racked by physical ailments but filled with unquenchable vitality, a man whose formal writing tended toward the heavy and the abstract but whose conversation and writings both were studded with crisp aphorisms and deft turns of phrase, a man capable of making precise philosophical distinctions but also prone to utter rash statements in the heat of argument, a proud and self-assertive man who could be sensitive to others and could inspire deep loyalty and lasting friendships. He would be, in short, a very gifted and very *human* human being.

But if Ascoli shares these traits with a man of another nation and another era, he must also be seen as the product of his own nation and his own times: as a follower of the Italian liberal philosopher Benedetto Croce, as an opponent of Benito Mussolini from the very beginnings of Fascism, and as a voluntary exile from Italy who chose to become an American citizen. "It did not cause me any trouble to become an Italian," Ascoli once wrote; "but my becoming an American is my own work."[4] Born in Ferrara on June 25, 1898, the only child of Enrico and Adriana Ascoli, he spent his first thirty-three years in Italy. Because of his extremely delicate eyesight—he began wearing glasses at the age of four—he had a somewhat protected childhood: the doctors ruled out all athletics—running, bicycling, swimming—and advised against fast or bumpy means of transportation. When his father finally bought a car—"after having consulted God knows how many ophthalmologists in the Po Valley region"—the chauffeur was given strict orders to drive no faster than twenty kilometers an hour.[5] Despite the trouble with his eyes—his fellow students sometimes had to read aloud to him—he earned his LL.D. at the University of Ferrara in 1920 and taught legal philosophy at several Italian universities before receiving his Ph.D. in philosophy from the University of Rome in 1928.[6] His first book, a brief study of Georges Sorel, appeared in both Italian and French editions in 1921.

Having opposed Fascism from its beginnings in October 1922, Ascoli found that the only magazines in which he could publish were anti-Fascist publications:

> From the March on Rome to the beginning of totalitarianism in January 1925, I had been writing for *Rivoluzione Liberale, Il Quarto Stato,* and a few other anti-Fascist publications. From January 1925 on, I contributed also to the first underground sheet in Italy, *Non Mollare,* but ceased to write for it after a man in Florence whom the Fascists considered the writer of an article of mine was murdered.[7]

In 1928 Ascoli was arrested by the Fascist police. Talking some forty years later about this experience, Ascoli said:

> When I was against Fascism it was an entirely voluntary position, different from Germany, where a certain category of people were eliminated immediately—Jews. All the trouble I had with Fascism was because of what I had written and most of all because I would *refuse* to enter the Fascist Party. Even when I was put in jail—and I thought they had caught me for something I had really done—at the end they told me, "Professor, we need brilliant people. Why do you remain in such a stupid position? It's as if you're not alive—you can't publish anything." So . . . it was an act of will that imposed upon me a sort of isolation and sterility. Very unpleasant.

Three weeks in jail were followed by a period of modified house arrest. For the next three years he was shadowed by the police, but as he later wrote: "The company of the police I resented the least: Fascism was a totalitarian system tempered by sloppiness."[8] In the fall of 1931 all university professors in Italy were ordered to take an oath of allegiance to Mussolini and the Fascist regime. Only twelve refused.[9] A few others, like Ascoli, managed to get out of the country.

"When the Rockefeller Foundation almost miraculously gave me a fellowship, my name was in the black book. When I left Italy, I knew that I couldn't come back, unless I wanted to go to jail," Ascoli recalled forty years later. He landed in America on October 5, 1931. That remained an important date for him, he later wrote, because on that day nine years later he married his wife Marion, daughter of Julius Rosenwald, the Sears Roebuck executive and philanthropist.[10] His first two years in America were spent as a Rockefeller Fellow at Harvard, the University of Chicago, and the University of Wisconsin. "I realized very quickly that there was no market for legal philosophy. So it did not take me much time to change the label: instead of legal philosophy I called it political philosophy." In 1933, when Alvin Johnson, director of the New School for Social Research, set up the University in Exile—known more formally as the Graduate Faculty of Political and Social Science—he appointed Ascoli, "the most brilliant of the younger Italian political scientists," as professor of political philosophy.[11]

The University in Exile was originally conceived as a means of helping university professors who had escaped from Nazi Germany. As the lone Italian among its original eighteen members, Ascoli concluded that perhaps his parents had shown unusual foresight in naming him Max. After he became a citizen in 1939 he wrote, "I am in a large group of friends over here, a community which is growing out of the work of each one of us, and mutual confidence, and common beliefs."[12] Ascoli contributed frequent articles and book reviews to *Social Research,* the quarterly published by the Graduate Faculty; in 1937 he and Fritz Lehmann edited a collection of articles by their colleagues, *Political and Economic Democracy.* The following year he and another colleague, Arthur Feiler, collaborated on *Fascism for Whom?* From 1940 to 1941 he served as dean of the Graduate Faculty; in 1964 at a convocation marking the thirtieth anniversary of the University in Exile, he was one of four former faculty members to receive an honorary Doctor of Letters degree.[13]

Ascoli's first book in English, *Intelligence in Politics,* appeared in 1936. In the years before World War II he was also writing articles for the *Atlantic Monthly, Foreign Affairs, American Scholar, Yale Review,* and the *Annals of the American Academy.* As he later wrote: "The Fascist revolution struck me as a pathological degeneracy of democracy likely to ruin other nations besides Italy. . . . Most of my writing was on my pet obsession: how, in the democratic climate of our times, freedom can perish."[14] His first appearance in print in America had been in the *International Journal of Ethics* (January 1933) as one of six critics invited to comment on Charner M. Perry's article, "The Arbitrary as Basis for Rational Morality." Ascoli was identified as being with the University of Cagliari, and his brief comments on an abstract ethical question bear an interesting relationship to his actual situation there. Arguing that morality transcends law because it is more universal, he cites a hypothetical case of his consciously disregarding the laws of his city. "The law in this case is only the occasion of my action," he says; "the punishment is the price of my responsibility; there is no system on earth liberal enough to make free or cheap any radical opposition."[15] What he does not add is that the hypothetical case was an actual one: at the University of Cagliari he was the only faculty member who was not a member of the Fascist Association of University Professors and also "the only faculty member who had his own personal policeman."[16]

In an *Atlantic Monthly* article that same year, "Fascism in the Making," he strongly refutes the notion that Roosevelt's New Deal could be the first stage of an American Fascism. He lists six characteristics of Fascism: (1) the emergence of "small bourgeois desperados" as self-appointed leaders; (2) the revolt of youth against age; (3) the setting up of the state as an object of idolatry; (4) the prohibition of politics; (5) the transformation of Fascist politics into a kind of religion; and (6) "a peculiar kind of intellectual disintegration . . . a corrosive and debauching criticism of fundamental moral and political beliefs . . . a nihilism that brushes reason aside and insists that the only realities are 'cold facts and successful deeds' . . . a despairing and disparaging view of ordinary human nature and an absolute faith in the impersonal forces of history." Ascoli sees the United States as possessing none of these—except the last—in any but small and homeopathic doses.[17] The last characteristic is worth noting, for it remained one of Ascoli's most persistent concerns, from his study of Sorel in 1921 to an article on campus riots he submitted to the *Wall Street Journal* in 1969.[18]

A second article in the *Atlantic Monthly* contains one of Ascoli's earliest comments on American liberalism, which he sees as depending not so much on the "professional liberal"—"the smaller the number of priests in any religion, the better"—as on the "conditions or attitudes deeply rooted in the structure of society" of America.[19] By 1937 he was disenchanted enough with some aspects of Roosevelt's program to write his first article on an American domestic controversy, an article for the *New York Herald Tribune* attacking Roosevelt's attempt to "pack" the Supreme Court.[20]

Some of Ascoli's most important and persistent ideas on political theory are contained in two rather controversial articles he wrote at the end of the 1930s, "The Right to Work,"[21] and "Freedom of Speech."[22] In the former he argues that rights "*do* have a cost. . . . The distinction between effective, enforceable rights and demogogic wishes can be established mainly in terms of what the citizens are able and willing to pay." Rights require efforts, duty, and work; the guarantee of rights by society to the individual "is related to the essential principle that the realization of human personality is the greatest goal in life."[23] The right to work, he was to argue later, appears to be the foundation of all other rights, for "we know freedom when we do our work well."[24] But the right to work does not mean automatic and permanent certainty of a job; it means only the "right to all possible assistance in finding a job."[25] For a democracy cannot abolish unemployment; only a totalitarian society can do that.[26] Making an analogy with the limits of medicine, Ascoli argues that there are limits to what a government can do. "Life is a struggle," he maintains, "and the course of civilization has been a constant attempt to subject that struggle to rules."[27] Amplifying his argument later in *The Power of Freedom*, he describes the basic dilemma involved in any discussion of the right to work:

> Civilization can hardly survive unless the large masses of men who have sunk to the lowest level of freedom are given the opportunity to free themselves from want and from fear. Yet no organization, whether national or supernational, can become the monopolistic giver of work to its citizenry without ultimately threatening the basis of their humanity. Any government which attempts to take over the function of freeing men from want and from fear is likely to become the main cause of want and the object of the most hopeless fear.[28]

His article "Freedom of Speech" insists that freedom of speech is not an absolute but a means to an end: "Freedom of speech, to have social meaning, as opposed to magic meaning, demands leadership and sufficient personal examples of what it should be used for. . . . Free speech, like every other social instrument, demands the energetic application of standards if it is to fulfill a human need."[29] Having seen the Italian press under Mussolini enjoy "the freedom of talking itself into subjection,"[30] Ascoli attacks the "neo-liberals"—whom he does not otherwise identify—who seem to advocate a laissez-faire in ideas, "an eternal grant of immunity and indulgence to every uttered or utterable word.[31] Rather, free speech should be seen as he believes the Founding Fathers saw it, "as a weapon against oppression, as a tool for finding out the truth in any given situation." And "the whole assumption in free speech," he argues, "is that men can be governed by reason."[32] Thus he maintains, as Milton had in his "Areopagitica," that there are definite limits to tolerance: "The liberal faith leads to intransigence toward illiberal ideas and to charity toward illiberal men. . . . Those who believe in it and practice it have to be equally skilled in the art of mercy and in the handling of the whip."[33]

Much of Ascoli's writing in the thirties and forties, naturally, concerned Italy. Early in 1936 he joined G. A. Borgese, Michele Cantarella, Guido Ferrando, Gaetano Salvemini, and Lionello Venturi in a letter to the *New York Times* attacking the myth that Mussolini had saved Italy and Europe from Communism. The six writers proposed a noncommunist "social revolution" for Italy.[34] Ascoli had been one of the small first wave of anti-Fascists to arrive from Italy; the majority of them did not come until after the passage of racial and other restrictive laws in 1938.[35] In 1940 he was selected as the first president of the Mazzini Society, an Italo-American group founded to oppose Fascism and every other kind of dictatorship. Salvemini was named honorary president, and Alberto Tarchiani, who was to become the Italian ambassador to Washington early in 1945, was named national secretary.[36] Together with Borgese, Salvemini, and Count Carlo Sforza, Ascoli called upon the Mazzini Society, at a meeting attended by 1,500 people in February 1941, to reject "all totalitarian ideologies."[37]

During World War II and shortly afterward Ascoli wrote articles on Italy which appeared in *Free World*, *Commonweal*, and the *Journal of Politics*. In 1948 he edited, with a long critical introduction, *The Fall of Mussolini*, a translation of Mussolini's own story that the deposed dictator had written for Italian newspapers shortly before his death. Ascoli's concern for conditions in postwar Italy led him in 1944 to form a nonprofit organization, the Committee for Assistance and Distribution of

Materials to Artisans (CADMA). A practical application of the theories propounded in "The Right to Work," CADMA set to work to revive the handicraft industries in Italy that had been destroyed by the war. CADMA was later merged with Compagnia Nationale Artigiana, an Italian government agency, and its American branch was discontinued in 1953. "This little pre-ECA project," Ascoli wrote, "did good work."[38]

Two other activities in the 1940s, one as a United States government official and the other as a journalist, also played a formative role in shaping the editor and publisher of *The Reporter.* In 1941 and 1942 Ascoli served as associate director of cultural relations in the Office of the Coordinator of Inter-American Affairs, which was headed by Nelson Rockefeller. Recalling this experience thirty years later, Ascoli said:

> While I was traveling through Latin America as an agent of good will and of discreet information for Nelson Rockefeller and above all for the State Department, I was *appalled* by the character of some of our military or civilian missions in the Latin American countries—their chumminess with sometimes unclean, rather dubious leaders. Their tendency, as I remember I wrote for the government at that time, was either to remain cloistered in a group of whiskey-drinking Americans or to be imitation natives. Before we Americans—I say Americans because I have become as you know an American citizen in the full sense of the term—before we learn, how many Vietnams are we going to need?

Ascoli's journalistic experiences included an association with *Free World*, a monthly magazine which first appeared in October 1941 and continued through 1946, when it was merged into the newly formed *United Nations World.* Founded as a self-described "political act" under the directorship of Clark M. Eichelberger, Freda Kirchwey, and Simon Marcovici Cleja, *Free World* espoused antifascism as its raison d'être: "Democracy and fascism cannot live together. . . . In the society for which we fight there is no place for any form of fascism."[39] The nature of *Free World*'s antifascism emerges most clearly, perhaps, in two of the eight points of its official program:

> 5. Freedom of travel . . . (. . . This provision, of course, excludes fascists from the freedom of travel because their program stands for murder and torture.)

7. Complete destruction not only of the Axis regimes, but of fascist and nazi ideas everywhere.[40]

Leaving aside the question of how the "complete destruction" of any idea is to be achieved, those who have grown accustomed to hearing "free world" as a cliché of anticommunist rhetoric may find a certain illumination in substituting the words "communism" and "communist" for "fascism" and "fascist" in *Free World*'s statements of principle.

Early issues of *Free World* listed a seventy-three member International Honorary Board that included such diverse figures as Ascoli, Robert Bendiner, Eduard Benes, Madame Chiang Kai-shek, Albert Einstein, Harold Ickes, Max Lerner, Archibald MacLeish, Gunnar Myrdal, Reinhold Niebuhr, Ralph Barton Perry, T. V. Soong, Clarence Streit, and Dorothy Thompson. Several articles by Ascoli on Italy and on the San Francisco Conference appeared in the magazine in 1944 and 1945. In September 1945, he was named as one of *Free World*'s five editorial writers, joining William L. Shirer, Orson Welles, Lin Yutang, and Norman Angell. His last appearance in the magazine was as a member of a round table discussion, "World Government vs. the United Nations," in the summer of 1946. Here he showed himself a strong supporter of the United Nations as opposed to what he considered the utopian schemes of world government advocates. "Peace is not a condition of rest or perfection," he argued. "It is simply a state of affairs where the conflicts among nations are fought with weapons that are those of politics and not those of military warfare."[41] Thus peace cannot be produced by world government but by a "new pattern of human relationships where the satisfaction of universal pre-political needs exerts a check on political or power contentions"—a pattern he saw as beginning to emerge in connection with the United Nations.[42]

This insistence on "pre-political needs" is characteristic of his approach to social problems. Both as a political scientist and as a journalist Ascoli insisted that politics is only a part of human affairs—in sharp contrast to those, including many liberals, who see life as wholly political. As he expressed his view most epigrammatically, "The individual is sovereign in the election booth, but no man can spend his life in an election booth."[43] And for this reason he could characterize Italian Fascism as "the mysticism of the purely political mind."[44]

Many of the ideas Ascoli expressed in his writings before 1949 reappear in his editorials for *The Reporter.* In their most concentrated—and often most abstract—form they may be found in his short book, *The Power of Freedom,* published in the year he started *The Reporter.* In the introduction to the book he lists the chief influences on his thinking:

> Aristotle's thinking has certainly influenced me more than Plato's, and the stoic over any Greek-Roman school of thought. Perhaps my most conscious indebtedness is to Italian political philosophy: Dante, Machiavelli, and above all, Vico, to whose thinking I have been exposed so thoroughly and unreservedly for so many years that I constantly realize how I follow Vico even when I think I have forgotten him. To some extent, this applies to Croce, too. . . .
>
> There is no hiding the fact that I love Burke and loathe Rousseau. Among American political thinkers, the writers of *The Federalist* are high at the top, with Hamilton second to none. When it comes to nineteenth-century social reformers, there is no love lost for Marx and Engels, while Proudhon and Sorel stirred me early in life. Of the political writers of the last century, the one whose work I have most admired and cherished is De Tocqueville.[45]

After giving this list, Ascoli labels himself "a liberal, and I don't want to add any qualifying adjective."[46] Commenting in 1971 on this list, he conceded that it includes "some people who are not exactly considered among liberals," and added that he had always been "a liberal in the old European tradition, and in a way I have always been, as a liberal, rather untypical as compared to the traditional types of American liberal." He described this "traditional" type of liberal as "always inclined to solve problems by the intervention of government," an inclination that profoundly disturbed him.

The insistence that the power of politics to produce human happiness is necessarily limited is a central idea in Tocqueville,[47] and one can find a number of parallels between Ascoli's first book in English, *Intelligence in Politics,* and *Democracy in America.* Both Tocqueville and Ascoli are products of European culture, trying to analyze that large and baffling phenomenon, American democracy. Both see democracy as having a kind of historical inevitability: "Tocqueville was right one hundred years ago: there are no ways out of democracy in our time."[48] Where Tocqueville sees "political association" as one of the guarantees of liberty, Ascoli sees "groups," including "ethnical groups," as a guarantee of individual freedom.[49] Like Tocqueville, Ascoli puts his finger on one of the central paradoxes of democracy: "Democracy seems a crowning point of history, and at the same time it makes every tradition it absorbs flat and shabby." It "offers to every seed a convenient hotbed where it can rapidly mature in unsavory but highly salable fruits." It "perhaps means the setting free of demos, and demos once free can make the use he pleases of his freedom."[50] (One can translate this idea from an abstraction into concrete terms, perhaps, by recalling the crowd behavior at Andrew Jackson's Inaugural, the vulgarities and prejudices of populism, or even the kinds of magazine covers displayed in the typical modern supermarket.) Yet, because American democracy has produced "the plain enlarged picture of man . . . it is difficult to condemn this democracy without condemning the race."[51]

Like Tocqueville, Ascoli is sensitive to the peculiar affinities between democracy and dictatorship. Tocqueville says, "Despotism, then, which is at all times dangerous, is more particularly to be feared in democratic ages"; Ascoli says, "The so-called ways out from democracy—bolshevism or fascism—are simply reënforcements of it, potentialities that every democracy has in its bosom and which explode when the social bottom is reached too quickly."[52] Thus Ascoli can describe fascism succinctly as "totalitarian democracy," "democracy gone blind," or "democracy without freedom."[53]

For Ascoli as for Tocqueville the only effectual remedy for the evils that may be produced by democracy is political freedom.[54] Freedom is the leitmotif of all of Ascoli's writing: "I believe that freedom . . . is a propulsive power of civilization—a power that . . . can drive the men of our time to goals so high and so good that we can only dimly discern them."[55] This deep and almost mystical belief in freedom probably best explains Ascoli's persistent description of himself as a liberal: he remembers the word's etymology.

Ascoli's indebtedness to Tocqueville can be more easily documented than his indebtedness to Croce, of whom Ascoli wrote in 1949, "I think I have been constantly moving away from his influence, but as a grown-up man moves away from his family."[56] The year Ascoli graduated from the University of Ferrara, he named Sorel and Croce as the two greatest intellectual influences on postwar youth, not because they promised a magic formula to solve the problems that youth faced, but because they communicated the strength with which they had solved their own problems.[57] But later this enthusiasm for Croce was to be tempered:

> This man, who offered criteria for discrimination in all things, possibly never thought that it would have been most important to stress the difference between right and wrong. At the core of his philosophy there was a certain moral vacuum, but the danger of this vacuum was not apparent since Croce himself was an extraordinary example of moral conduct.[58]

Ascoli said in 1971 that this moral vacuum, which led to Croce's support of Mussolini in the first days of Fascism, resulted from

> his Hegelianism—the justification of what exists because it exists. It happens; therefore it has to happen, and it is the best thing that could have happened; and he went on saying that about Fascism until the Matteotti murder. I remember a *terrible* quarrel I had with Croce in the days of the March on Rome. We quarrelled; and we were great friends after the war because I sent him some coffee.[59]

Yet Croce's intellectual influence on Ascoli persisted, and at least one sentence Ascoli wrote about him could be applied with equal accuracy to himself: "Always a liberal, he was impatient of the sanctimonious style of the priests of liberalism."[60] Croce's historical writings have been described as "a sustained warning to Europe against the dangers of simplicism in thought and fanaticism in action in their constant repetition of the conviction that even the highest ideals become arid and oppressive when they are pushed to the outer limits of their potential development." Thus Croce was convinced that the hallmark of those who understood the modern conception of freedom was their "rejection of every hope for a final solution to human problems."[61] Here is clearly one source for Ascoli's impatience with simplistic slogans and his skepticism toward some liberals' utopian dreams.

To Croce's distrust of final solutions, Ascoli adds a strong distrust of intellectuals, who "know better than any other group how to enjoy a civilization and how to undermine it."[62] He had seen an unforgettable example of intellectual betrayal in the career of Giovanni Gentile, a friend of Croce and one of the leading Hegelian philosophers of Italy, who became Mussolini's first Minister of Education in order to institute needed educational reforms but who sold himself to Fascism in the process. Gentile's example, Ascoli later wrote, showed that "there is no greater danger for the intellectual than to accept the temptations of political power," for it is a betrayal of intellectual freedom to be "too clever, too Machiavellian, accepting whatever means is offered to us to reach our goals."[63]

What Ascoli had seen in Italy also gave him an abiding suspicion of the welfare state, for Fascism could be described as "liberalism without freedom":

> In a way Fascism did have a liberal function. The coalition government under the Fascist leadership rapidly achieved goals for which the liberals had long striven. In ten months it enforced some of the most radical reforms that liberal Italian educators had vainly proposed for twenty years. It balanced the budget. It managed to reduce taxation. It vigorously pursued the program of social legislation and low-cost housing which the previous democratic administration had started.

Thus the young Fascist leaders, "busily at work learning their new job as rulers, could think of themselves as the knights of liberalism."[64]

In an early issue of *The Reporter,* at a time when there was much talk of a noncommunist affidavit, Ascoli proposed the following nonfascist affidavit:

> I do not believe in bigness. I think that big labor, big business, and big government can and must be checked. I believe that when the three bignesses get too chummy, one helping the other to get bigger, then the danger of Fascism is upon us. I pledge myself to uphold the fullest cooperation of the competitive system in business, in labor, in politics.[65]

A few months later he suggested that the problem of the relationship among these three "bignesses"

> can find its solution in the international setting in which America now operates. . . . Actually, the challenge that now confronts the leaders of American government, business, and labor is to skip the phase of imperialism and of colonial exploitation, while improving the level of productivity and well-being in foreign countries.
>
> Perhaps this is the ground where the various brands of liberal-

> ism championed by the leaders of government, business, and labor can find their point of convergence.[66]

There is a logical progression in Ascoli's thought. An Italian immigrant whose supreme value was freedom, he felt that he had found the potentialities of freedom most fully realized in America. He had at the same time seen at first hand some of the shortcomings of American representation abroad; he had been associated with the internationally minded *Free World;* and he was grappling with the problem of "bigness" in American life—what a younger generation of social scientists would attack as the System, the Establishment, or the Power Elite. His experiences and his thinking were to culminate logically in the conviction that was central to *The Reporter*: "We are in the business of world politics for keeps" because "the American experience cannot in any way remain the peculiar and exclusive privilege of the American people."[67]

Yet neither this nor any of Ascoli's other convictions can be fully appreciated without coming to terms with a somewhat hidden factor in his thinking—his religious beliefs. In his "Farewell to Our Readers" in the final issue of *The Reporter* he wrote:

> Some of the major *Reporter* failures came from its being overrestrained by the fear of being predictable and of sounding preacherish. Perhaps one of the most grievous mistakes has been mine: I muzzled myself when it came to spiritual and religious matters because I was afraid of showing myself for what I am: a deeply religious man.[68]

Looking back more than two years after *The Reporter* folded, however, Ascoli was not sure what he could have done differently. As a Jew who had long been profoundly captured by the figure of Christ, he understandably felt the need for reticence: "Jew" and "Christian" have for so many centuries been regarded as mutually exclusive terms that his views presumably would not fit into any popularly accepted category.[69]

But his religious views found some expression nearly every year in his editorials for the Christmas issues of *The Reporter*:

> Our modern civilization is so deeply rooted in Christianity that it can sometimes afford not to consider itself Christian—particularly since the patterns it has created have been so eagerly adopted by

> peoples untouched by the Christian experience. Yet the nations that still call themselves Christian should not be allowed to forget that the ideals they live by are more or less veiled translations or secularizations of Christian principles. Perhaps these principles had better remain translated and secularized to avoid the curse of clericalism and theological dispute. But how can we be oblivious—at least at Christmas time—of the fact that the principle of freedom, sometimes called liberalism, is nothing but a translation in abstract terms of the Christian idea of man as the temporary responsible bearer of divine creativeness?[70]
>
> If Christ or the legend of Christ had never come into the world, men would have found some other way to celebrate Christmas. For Christmas means freedom from the chain of necessity, from the obligation of past deeds and misdeeds that keep us bound.[71]

At the same time Ascoli had an intense dislike of anything that sounded "churchy," and this dislike manifested itself continually in *The Reporter,* where both Ascoli and his staff regularly attacked the unctuous piosity of politicians—Eisenhower and John Foster Dulles were favorite targets—and anything smacking of government-endorsed or commercialized religion.[72] "I *hate* religiosity," Ascoli said in 1971, adding, "I think, like my friend Ignazio Silone, who in his simple way has said such a profound thing, that the most important date in history was December 25 approximately 1971 years ago. In that sense I am profoundly a Christian."[73]

In his study of Sorel in 1921 Ascoli had expressed a view later to be stated by Croce in his famous essay "Perche non possiamo non dirce 'cristiani' " ("Why We Cannot Help Calling Ourselves Christians"), that the rise of Christianity was, in Croce's words, "the greatest revolution that the human race has ever accomplished."[74] Thus there is at the heart of Ascoli's thinking an element that seems to distinguish it from the classic Italian liberalism as described by Croce, liberalism which "lacked the support of a religious creed such as it had enjoyed at other times and in other countries."[75] Croce's own teaching, Ascoli felt, failed to give his followers what they needed to meet the pressure of Fascism: "The intellectuals trained by Croce . . . needed . . . definite criteria of right and wrong and active faith in liberalism and an immediate link with the tradition of the Risorgimento."[76] Ascoli, however, never attributed his own early opposition to Fascism to anything so formal as a religious creed or set of intellectual principles. Looking back years later, he could simply say, "At a certain moment your own conscience makes you say, 'No!' "

During the last seven or eight years of *The Reporter,* Ascoli was accused of turning "conservative," but he argued—justifiably—that he had been consistent in his views over the years. When he was attacked for his editorials on the Berlin Wall, he wrote in a piece headed "Somewhat Personal":

> They can call me what they please. I have been a liberal all my life. I know that political freedom is a very difficult thing to organize and make operational, but I know it can be done. It has been done in this country, of which I am proud to have become a citizen. But I do feel that my own freedom is both curtailed and endangered by the slavery of peoples in other lands.[77]

In the 1960s Ascoli strongly attacked the new isolationism among liberals and named as the "major advocate" of that isolationism a man who had become something of a hero to many liberals, Walter Lippmann.[78] His disagreement with Lippmann on this issue was hardly new. Thirty years earlier he had reviewed Lippmann's *Method of Freedom* and criticized it for what Ascoli felt was Lippmann's belief that liberalism was best suited to Anglo-Saxon countries, while other political philosophies were suited to other nations. This, Ascoli had written, was "a very pernicious form of egoistic nationalism. . . . Political beliefs do not have much chance to survive if they do not take the risk of being universal."[79]

Continuing to maintain that his stands on political issues were matters of conscience, Ascoli wrote a second piece headed "Somewhat Personal" early in 1968, at a time when he felt that certain politicians and intellectuals—privately he called them "profiteers of chaos"—were trying to lead the world either to fascism or to "the permanent hell of anarchy." He not only argued for his consistency but expressed a certain anguish:

> After having exercised my capacity for dissent in two countries, against two men [Mussolini and Senator Joseph McCarthy], both of whom turned out to be charlatans, I thought I could live a more or less peaceful life as a man deeply concerned with ideas

> and politics but obsessed by neither. And now I am going through the third round. This time, it all comes from the fact that I refuse to join the dissenters. . . .
>
> I must confess that this last bout is a singularly bitter one. On the other side, there is a mob that calls itself liberal and in that mob there are people who have been and are dear to me. A has-been friend is a horrible thing that I still refuse to accept. The feeling of loneliness I have experienced twice sometimes returns. Yet I must say that I have no qualms. This does not come from the fact that I have been proven right twice. Rather, it comes from the conviction that when dissent is so pampered by public officials and is so freely used, it is worth exactly what it costs to be exercised: nothing.
>
> My dissent from the dissenters is neither easy nor cheap, and if I am wrong, I can only say that I cannot do otherwise.[80]

Although Ascoli's support of President Johnson's Vietnam politics happened to be the occasion of the final split between him and a large body of liberals, signs of divergence can be detected as early as the midfifties. But they went largely unnoticed, for a reason suggested by one former staff member:

> The time when Joseph R. McCarthy was doing what he was doing, someone who was speaking out strongly against Joseph R. McCarthy, which Max Ascoli certainly did, was obviously persona grata among all people who considered themselves liberals or anti-totalitarians or whatever you will. So the distinctions weren't noticed then that I think were there all along. I would support Ascoli to a large extent that he didn't change—as much at least as people claim he did.

One of the first clear signs of Ascoli's divergence from some elements of American liberalism was the unexpected quarrel that developed between him and Arthur Schlesinger, Jr., in the spring of 1956. Several former staff members consider this quarrel to have been a major turning point in the relations between *The Reporter* and some important elements of the American liberal community. For Schlesinger, who had met Ascoli after the war, had been consulted about the magazine and its fledg-

ling stage and had helped recruit one of its senior editors. A number of his articles, including a prepublication excerpt from *The Vital Center,* had appeared in the early issues.

The quarrel began with a friendly and rather lighthearted exchange of letters. In December 1955, Schlesinger sent Ascoli a memorandum on liberalism and asked for his comments on it, suggesting that it might serve as the basis of an article for *The Reporter.* Ascoli replied that he had many comments on the memorandum, including some disagreements. Instead of exchanging memos, Ascoli suggested, "Why don't we make this thing a sort of public debate?" He thought a debate would be both fun and useful. "There is no other friend and fellow liberal with whom I would like better to have a sort of family debate on American liberalism today," he added. Ascoli proposed that Schlesinger submit to *The Reporter* an article based on his memorandum, and the article would be published exactly as written. Ascoli would then publish a piece in answer to it in the same issue. (Ascoli had done something similar in 1950 with Isaac Deutscher.) "Should you then wish to have a second round in a later issue," Ascoli continued, "of course with shorter articles from both of us, you have only to say the word."[81] Schlesinger agreed and submitted his article a few days later.

The two articles appeared together in the issue of May 3, 1956, as "The Future of Liberalism—a Debate." Schlesinger's article, entitled "The Challenge of Abundance," argued that it was time for a shift from the "quantitative" liberalism of the 1930s to a "qualitative" liberalism that would aim to deal with the current "inner unrest" and "widespread discontent of some kind":

> Today we dwell in the economy of abundance—and our spiritual malaise seems greater than ever before. As a nation, the richer we grow, the more tense, insecure, and unhappy we seem to become. Yet too much of our liberal thought is still mired in the issues, the attitudes, and the rallying cries of the 1930's.

Thus, Schlesinger argues, "What is required today is a new liberalism, addressed to the miseries of an age of abundance." Toward the end of the four-page article are two paragraphs on foreign policy which argue that "a truly creative and progressive American foreign policy can only come from a truly creative and progressive America" and continue:

> We will thus probably require a reawakening of the liberal conscience and the liberal will at home before we can offer positive and compelling alternatives to the world. We cannot convincingly champion freedom before the world so long as we kick freedom around at home. We cannot convincingly champion equality abroad so long as we practice segregation at home. We cannot convincingly champion opportunity abroad when too many of our own people linger at home in cultural mediocrity and economic want. As we renew a fighting faith against the inequities of our own society, we will generate an enthusiasm that will reverberate across the world.

Schlesinger is vague as to how these desirable ends are to be achieved, but clearly he thinks domestic policy should have priority.

Ascoli in his rebuttal, entitled "The Scarcity of Ideas," accuses Schlesinger of basing his "comfortable, cozy" liberalism on the assumption that "the progressive and bettering job can best be done by government taxing and spending"—provided that "enough liberals hold positions of power" in government. Ascoli argues: "American liberalism must acquire a far greater sophistication toward power and learn how not to hate and not to love it. The liberals' anti-business demonology is about as outdated as their—alas—frequently platonic love of government." But, he admits,

> what really stirred me to pick a quarrel with Schlesinger was the offhand way in which he brushed off the impact of international affairs on the fortunes, well-being, and freedom of our country. Maybe what's new and surprising is not the new phase of liberalism he advocates but the rebirth of a virtuous "I-am-unholier-than-thou" isolationism. The idea that before getting messed up in other people's business, we ought to put our own house in order and realize in all its fullness the American dream is an old boiler plate of liberal rhetoric.

The trouble with Schlesinger's argument, says Ascoli, is that the world will not stand still while we spruce up our domestic life. Moreover, he sees a continual interplay between foreign and domestic policy: American progress toward a better society at home has often been a by-product of its involvement in world affairs:

The conclusion, at least for liberals, should be that they must enter upon their task of reversing the trend and making freedom operational both at home and abroad—for there is no line of demarcation between home and abroad.

In the issue of May 31 the debate continued, with Seymour E. Harris, Thomas K. Finletter, A. A. Berle, Jacob Javits, and Leon H. Keyserling as additional participants. Harris and Finletter appear to be committed partisans: the former defends Schlesinger's suspicions of big business; the latter maintains that Ascoli has ignored Schlesinger's impeccable credentials as an internationalist. Berle, a personal friend of Ascoli's, tries to act as peacemaker and stresses the points of agreement between the two debaters. Javits' brief statement ignores the main points of the debate and merely takes issue, as a liberal Republican, with Schlesinger's apparent identification of liberalism with the New Deal and Fair Deal elements of the Democratic Party. Keyserling seems generally more critical of Schlesinger than of Ascoli. "Everyone is for God, Home, and Mother, and 'qualitative' liberalism impresses me the same way," he writes, adding, "It seems to me that Schlesinger claims too much for a 'liberal' government, and that liberals can better strengthen their cause by facing with humility problems which even they thus far have been unable to solve."

Schlesinger's contribution to the continued debate—he refers to it as "this so-called debate"—is largely a personal attack on Ascoli, whom he accuses of being irrelevant and irresponsible, and of contriving "many distortions . . . not only of my views but of the general position and philosophy of American liberalism." In the final article of the debate Ascoli alludes sadly to the personal tone of Schlesinger's reply, although there are personal elements in his own statement.

The debate caused Schlesinger to sever his relationship with *The Reporter* and with Ascoli personally, even though mutual friends tried repeatedly to heal the breach. "It lost us a lot of friends," says one former staff member. "It did alienate people like John Kenneth Galbraith and certain others," says another. Galbraith, however, continued his personal friendship with Ascoli and continued to write for *The Reporter* until early in 1966. But he confined himself largely to book reviews, reserving his more serious articles on economics for other magazines. This, one former staff member speculates, may have been his way of showing his feelings about the affair. More than ten years after the debate Ascoli felt it "somewhat unnecessary and most unpleasant, but nevertheless proper" to name Schlesinger and Galbraith in an editorial as two of the power-hungry intellectuals Ascoli believed were responsible for "the present disarray" in the country.[82]

Ascoli said in 1971 that at the time of the debate one of his close friends, Henry Kissinger, complained to him privately, "How is it possible that you can accuse Arthur of being isolationist?" But from the perspective of the 1970s his criticism of Schlesinger appears more perceptive than it may have appeared to many at the time. In 1956 their differences on foreign policy seemed largely a matter of emphasis—a real disagreement though not necessarily an irreconcilable one. But the whole national debate over Vietnam in the 1960s can be seen, in one respect, as a polarization between the positions taken by Schlesinger and by Ascoli in the mid-1950s. Should America turn inward and solve its own problems first, or could it solve its serious internal problems only as it remained actively involved in world affairs?

At the time, of course, the debate represented a clash between two proud and abrasive personalities, and each man may have somewhat misrepresented his opponent's arguments, as is customary on such occasions. The true cause of their quarrel, however, may lie in something that was not explicitly stated on either side: a fundamental difference in the two men's conception of human nature and of the role of politics in human affairs. Schlesinger seems to regard man as a wholly political animal. And although he has had sharp disagreements with political radicals, he often seems to argue from a premise that is similar to theirs: namely, that the only effectual approach to the betterment of society is an approach that is political and institutional.[83] Thus he can maintain in the debate: "What liberalism requires is a program sharply focused to meet the qualitative discontents of the present age." This statement would appear as nonsense to those like Ascoli who believe that it is precisely the "qualitative" discontents—what Schlesinger had referred to in the debate as "spiritual malaise"—that no *political* program can meet.[84] Ascoli sees man as only partly a political entity; he is a spiritual entity as well. Schlesinger's arguments represent what often seems to be a dominant aspect of modern liberalism—its insistence on being pragmatic, rationalistic, and secular; Ascoli's arguments, by contrast, reflect an aspect of liberalism that is often hidden or submerged—its recognition of spiritual or transcendent values. Given such different frames of reference, it is no wonder that there was no meeting of minds.

Those who have felt and expressed the spiritual side of man's nature most strongly have not always—unfortunately—exhibited a saintlike humility or sweetness in their lives and writings. Milton and Dante come to mind as conspicuous examples of men who combine a deeply religious outlook with an all-too-human prickliness. Ascoli, too, has a prickly personality. When I told him during an interview in early 1971 that I had been talking with former members of his staff, he said: "Nobody must have spoken honestly to you without telling you that I am a difficult man to get along with. That when I am furious I am furious. But that also my mother used to say about me. That's part of my personality." Speaking of one staff member who "left in a huff," Ascoli added, "Well, I fired her. I was in a huff, too." (The two were later reconciled.) Speaking of the liberal friends who used to stay with him on visits to New York City but who stopped seeing him because of differences over Vietnam, he said, "It is a real tragedy. . . . On the whole I don't give a hoot in— it *hurts.* But I don't give a hoot in hell."

Ascoli's attitude toward members of his staff has been described by a few of them as "rude," "tyrannical," "abusive," or even "cruel." But at the same time he had an almost uncanny sensitivity to his associates' moods and problems. *The Reporter* had surprisingly little turnover of its editorial personnel during its nineteen years, and although some former staffers broke with him completely, probably at least as many—including some who disagreed with him over Vietnam—remained friendly toward him. His manner toward the women on his staff obviously charmed some of them greatly; the men were more likely to resent his manner. "I blew up at him a few times," says one former staffer; "and then I guess he respected me." The comments of another former staff member reflect an ambiguity that others may share:

> I've said a great deal that sounds as if I'm very critical of Ascoli, but I would certainly repeat that I have the greatest respect, affection, and *gratitude* to him not only for what he did for me, but for starting the magazine. . . . You know, he's one of those delightfully outrageous men. . . . No press lord ever had more presence or self-assurance than Max Ascoli. When the buzzer rang, I had to jump, I'll tell you. But here was this very human man. As I say, he'd sense that I was—maybe not even disagreeing with him, but some personal problem—he'd sense it, and

he'd be very quick to move in on it and see what he could do about it. He's someone who's enriched my life considerably.

One aspect of Ascoli's life that has bearing on both his personality and on *The Reporter* is his chronic ill-health. "For over fifty years," he has written, "people who cared about me were concerned about my eyes. I was living on borrowed sight."[85] He feels that his serious trouble with his eyes may have tended to make him an introvert. It was not until a successful operation for the removal of cataracts late in 1956 that he possessed anything resembling normal vision. His description of the operation shows the interplay between his physical ailments, his personality, and his political views:

> In the middle of November 1956, I entered The New York Hospital to be operated on by John McLean. He gave me a few warnings: my eyes might be improved but he could not guarantee it, or by how much. I must be prepared to be blinded for at least a few weeks, but it would help if I went on with my work as much as possible. Keep your mind on the move, he said. Perhaps the person who proved most helpful in not letting me get morbid about my eyesight was John Foster Dulles. That was the time of Suez, and I remember as if it were yesterday an editorial I dictated against the then-prevailing policy of passing the buck of major decisions to the U.N. . . . I was still in the hospital when I learned that at least another person, Hammarskjöld himself, shared my views of the leave-it-to-Dag policy.
>
> I don't remember ever having mentioned the Suez mess to McLean, but he had gotten into the habit of needling me: "How are we doing in the Middle East? What about the U.N.?" Invariably, I fell into the trap, and well after McLean had left the room I was still berating the insanity of siding with the Soviet Union against our two major allies, and the wrecking of the U.N.
>
> When the time for the first operation came, I felt self-possessed and somewhat detached.[86]

Ascoli's physical ailments were not confined to his eyes. About a year before *The Reporter* started, he fell and broke his hip. It did not set properly, and he had to undergo an operation a few months later. He had another hip operation—a risky one—the summer before *The Reporter* closed. "He just had one thing after another," a friend says. "There was always something—like Job, in a way. And he would joke about it, but I think it took a certain amount of courage." He once remarked to a staff member, "The only thing that hasn't been affected is my nose. That will come next."

His bodily ailments were accompanied by a certain agony of spirit; one friend describes him as "tortured." If one of his great strengths as an editor was his refusal to compromise with principle—the kind of principle that "may sometimes be very personal"[87]—a corresponding weakness may have been his indulgence in passionate enthusiasms and vehement hatreds. Whenever his public statements appeared to be lukewarm, there was likely to be a caldron of emotions boiling underneath. And sometimes he seemed tempted to take pride in his role as lone dissenter. As one former staff member puts it, "He hated to say what everybody else was saying, and I think this sometimes may have led him to say what everybody else was *not* saying, even if it did not happen to make very good sense."

Two earlier immigrants achieved fame as liberal journalists in America, E. L. Godkin and Joseph Pulitzer, and Ascoli shares some characteristics with each, such as the former's militancy and the latter's ill health. But unlike either of them he entered full-time journalism relatively late in life, after a long academic career. He seems a somewhat unlikely figure to have succeeded in the rough-and-tumble, opportunistic, and often amoral world of modern American journalism. If he was the catalyst that made his "experiment in adult journalism" a success for nineteen years, it is also obvious that it could not have succeeded "without the collaboration of the many distinguished American-born writers and editors on its staff or contributing to its pages or without the never-sufficiently-eulogized assistance of American money."[88]

NOTES

1. This and subsequent quotations from former staff members of *The Reporter* are taken from the author's interviews in 1971 with Donald Allan, George Bailey, Robert Bendiner, Robert Bingham, Harlan Cleveland, Philip Horton, Shirley Katzander, Irving Kristol, and Derek Morgan. Since some of those interviewed

did not wish to be quoted by name, sources for individual quotations in this chapter have not been indicated.

2. This and subsequent quotations from Ascoli not otherwise identified are from the author's interviews with him in January 1971.

3. He wrote a thoughtful account of his naturalization proceedings in "No. 38 Becomes a Citizen," *Atlantic Monthly,* February 1940, pp. 168-74.

4. Ibid., p. 170.

5. Ascoli, *Remembering Doctor McLean* (New York Hospital-Cornell Medical Center, n.d. [1971]). One of a series of pamphlets honoring John Milton McLean, surgeon-in-charge of the New York Hospital's ophthalmology division from 1948 until his death in 1968.

6. *Current Biography,* February 1954, p. 8.

7. Letter from Ascoli to the author, April 23, 1971. For a description of these publications, see Charles F. Delzell, *Mussolini's Enemies: The Italian Anti-Fascist Resistance* (Princeton: Princeton University Press, 1961), pp. 28-33. In January 1935, several years after he had come to the United States, Ascoli contributed an article to *Quaderni di "Giustizia e Libertá,"* a journal published in Paris by the anti-Fascist leader Carlo Rosselli. When Rosselli and his brother Nello were murdered by the Fascists in 1937, Ascoli paid them tribute in the *Nation* ("The Rosselli Brothers," July 3, 1937, pp. 10-11).

8. Ascoli, "Somewhat Personal," *The Reporter,* January 25, 1968, p. 10.

9. Ascoli, "The Fascisti's March on Scholarship," *American Scholar* 7 (Winter 1938), 57. Cf. Delzell, pp. 92-93.

10. *The Reporter,* December 23, 1952, p. 5.

11. Alvin Johnson, *Pioneer's Progress* (New York: Viking, 1952), p. 343.

12. "No. 38 Becomes a Citizen," p. 170.

13. *New York Times,* April 29, 1964, p. 46.

14. *The Reporter,* December 23, 1952, p. 5.

15. *International Journal of Ethics* 43 (January 1933), 156-57.

16. *Current Biography,* February 1954, p. 8.

17. "Fascism in the Making," *Atlantic Monthly,* November 1933, pp. 580-85.

18. Ascoli, "Campus Riots and the U. S. Government," *Wall Street Journal,* May 27, 1969, p. 22.

19. Ascoli, "Notes on Roosevelt's America," *Atlantic Monthly,* June 1934, pp. 654-64.

20. See *The Reporter,* December 23, 1952, p. 5.

21. *Social Research* 6 (May 1939), 255-68. The article, of course, was written before "right to work" became a slogan of the opponents of the union shop; it does not discuss the organization of labor at all, but simply the availability of work. Max Lerner wrote a five-page rebuttal to the article. See *Social Research* 6 (May 1939), pp. 275-80.

22. *American Scholar* 9 (Winter 1939-40), 97-110.

23. "The Right to Work," pp. 258-59.

24. Ascoli, *The Power of Freedom* (New York: Farrar, Straus, 1949), pp. 83, 88.

25. "The Right to Work," p. 267.

26. Ibid., pp. 260-61.

27. Ibid., p. 266.

28. *The Power of Freedom,* p. 89.

29. "Freedom of Speech," pp. 103-04.

30. Ascoli, "The Press and the Universities in Italy," *Annals of the American Academy* 200 (November 1938), 241.

31. "Freedom of Speech," p. 102. Thirty years later he was to call this the principle of "*loquor ergo sum* (I talk, therefore I exist)." *Wall Street Journal,* May 27, 1969, p. 22.

32. "Freedom of Speech," pp. 102-03.

33. Ibid., p. 110.

34. "Italy's Future: Free Country, Without Mussolini, Foreseen," *New York Times*, February 2, 1936, Section 4, p. 9.

35. Laura Fermi, *Illustrious Immigrants: The Intellectual Migration from Europe 1930-41* (Chicago and London: University of Chicago Press, 1968), pp. 117-18.

36. Fermi, p. 119; *New York Times,* February 3, 1945, p. 10.

37. "Fascism Assailed by Italians Here," *New York Times,* February 12, 1941, p. 13.

38. *Current Biography,* February 1954, p. 9. *New York Times,* August 2, 1945, p. 17; October 9, 1945, p. 4; July 31, 1947, p. 34; December 8, 1947, p. 42. *The Reporter,* December 23, 1952, p. 5.

39. *Free World,* October 1941, pp. 7-8.

40. *Free World,* January 1945, p. 50.

41. "Round Table No. 28: World Government versus the United Nations," *Free World,* June 1946, p. 23.

42. "Round Table No. 28, Part II: World Government versus the United Nations," *Free World,* July-August 1946, pp. 21-26.

43. *The Power of Freedom,* p. 123.

44. Max Ascoli and Arthur Feiler, *Fascism for Whom?* (New York: Norton, 1937), p. 317.

45. *The Power of Freedom,* pp. xii-xiii.

46. Ibid., p. xiii.

47. Jack Lively, *The Social and Political Thought of Alexis de Tocqueville* (Oxford: Clarendon Press, 1962), pp. 45-46, 65-66.

48. Ascoli, *Intelligence in Politics* (New York: Norton, 1936), p. 218.

49. Alexis de Tocqueville, *Democracy in America*, ed. Phillips Bradley (New York: Vintage, 1958), vol. 1, pp. 198-205; *Intelligence in Politics,* pp. 182-89.

50. *Intelligence in Politics,* pp. 207, 209.

51. Ibid., pp. 201-05.

52. Tocqueville, vol. 2, p. 109; *Intelligence in Politics,* p. 215.

53. *Intelligence in Politics,* p. 253; *The Reporter,* December 23, 1952, p. 5. The phrase "democracy gone blind" stems from one of Ascoli's earliest analyses of Fascism, an article titled "Il gigante cieco" ("The Blind Giant"), which he wrote in 1923 for Piero Gobetti's antifascist periodical, *Rivoluzione Liberale.*

54. Tocqueville, vol. 2, p. 113; *Intelligence in Politics,* p. 276.

55. *The Power of Freedom,* p. xiii.

56. Ibid.

57. Ascoli, *Georges Sorel* (Paris: Librairie Paul Delesalle, 1921), pp. 47-48.

58. "The Fascisti's March on Scholarship," p. 53.

59. Mussolini's March on Rome took place in October 1922. The murder of Giacomo Matteotti, an outspoken opponent of Mussolini, in June 1924, caused Croce and a number of other intellectuals to break decisively with the Fascist regime. See Delzell, pp. 12-15.

60. *Fascism for Whom?* p. 124.

61. Hayden V. White, "The Abiding Relevance of Croce's Idea of History," *Journal of Modern History* 35 (June 1963), 122.

62. *Intelligence in Politics,* p. 61.

63. Ascoli, "Education in Fascist Italy," *Social Research* 4 (September 1937), 347.

64. *Fascism for Whom?* pp. 44-47.

65. Ascoli, "The Non-Fascist Affidavit," *The Reporter,* July 5, 1949, p. 3.

66. Ascoli, "The Gilded Doghouse," *The Reporter,* October 25, 1949, p. 3.

67. Ascoli, "The Time Is Noon," *The Reporter,* November 8, 1949, pp. 2-4.

68. *The Reporter*, June 13, 1968, p. 18.

69. Ascoli's religious philosophy was set forth in a book he published in 1924, *Le vie dalla croce* ("The ways from the Cross"), a book that has never been translated and that is erroneously identified in the major biographical article about Ascoli—*Current Biography,* February 1954, pp. 7-9—as a biography of Croce.

70. Ascoli, "Christmas 1954," *The Reporter,* December 30, 1954, p. 7. Reprinted in *The Reporter Reader* (Garden City, N.Y.: Doubleday, 1956), pp. 13-17. With one paragraph omitted, it appeared as "Christmas" in *The Reporter,* December 29, 1966, pp. 10-11.

71. Ascoli, "Christmas 1959," *The Reporter,* December 12, 1959, p. 4.

72. See, e.g., Ascoli's praise for Churchill's disinclination to mix politics and religion ("Death and the Hero," *The Reporter,* February 11, 1965, p. 20); the attack in "Reporter's Notes" on *Life*'s spectacular "Christianity" issue ("Christianity with Scope," *The Reporter,* January 12, 1956, pp. 2-4); and many of the articles by staff member—and Protestant theologian—William Lee Miller, later reprinted in his *Piety Along the Potomac* (Boston: Houghton Mifflin, 1964).

73. Silone's statement may be found in "An Interview with Ignazio Silone," *Dissent,* Autumn 1954, p. 416.

74. Benedetto Croce, *My Philosophy* (London: Allen and Unwin, 1949), p. 37. Cf. Ascoli, *Georges Sorel,* p. 25.

75. Croce, *A History of Italy 1871-1915* (Oxford: Clarendon Press, 1929), p.253.

76. *Fascism for Whom?* p. 124.

77. *The Reporter,* November 23, 1961, p. 18.

78. Ascoli, "The Strange Case of Walter Lippmann," *The Reporter,* November 9, 1961, p. 24; "Separate Worlds," *The Reporter,* July 1, 1965, p. 12.

79. *Social Research* 1 (August 1934), 389-92.

80. *The Reporter,* January 25, 1968, p. 10.

81. Ascoli to Arthur Schlesinger, Jr., December 22, 1955.

82. Ascoli, "The Unpossessed," *The Reporter,* December 28, 1967, p. 15.

83. Schlesinger's approach is clearly stated in his first book, *Orestes A. Brownson:*

A Pilgrim's Progress (Boston: Little, Brown, 1939), pp. 38-40, and may be found in his subsequent writings. For a critique of this and some related aspects of Schlesinger's thinking, see Garry Wills, *Nixon Agonistes* (Boston: Houghton Mifflin, 1970), esp. pp. 336-44.

84. Thoreau, a moral radical, would be among those agreeing with Ascoli. So would Dr. Johnson, a conservative. See, e.g., the lines contributed by Johnson to Goldsmith's "The Traveller": "How small, of all that human hearts endure. That part which laws or kings can cause or cure!"

85. Ascoli, *Remembering Doctor McLean.*

86. Ibid.

87. Fermi, p. 283.

88. Ibid. p. 284.

Chapter Three

An Experiment in Adult Journalism

You see, it's a kind of game. Every two weeks the staff tries to put out a magazine, and every two weeks Dr. Ascoli tries to stop them. So far we've managed to win every time—but it's been close.
Attributed to an early managing editor of *The Reporter*

In a venture that was as risky as the one I undertook I have been very lucky. I had people who worked for me very loyally—no matter what an exaggerated conception they might have had of themselves.
Max Ascoli, 1971

During its nineteen years of existence, there could have been little doubt among those connected with *The Reporter* about who was the editor; a former staff member recalls editorial meetings as being like "a Byzantine court." Ascoli, looking back in 1971, described himself as an *editing* editor: "An editor must be something more than a ghost writer; he must also be a ghost thinker." One former staff member claims that Ascoli "tended to regard a manuscript as a tabula rasa on which he could put his own ideas, and some writers didn't go for that." But Jay Jacobs, a staff member who wrote nearly fifty bylined articles for the magazine, and who disagreed with Ascoli's views on Vietnam, insisted in

1968 that at no time had his copy for the magazine been messed about with.[1] Marya Mannes, who also disagreed with Ascoli over Vietnam, wrote in 1971 that Ascoli "was not only a very good editor but the only editor who has ever given me the freedom, the range, and the confidence to write as I chose and to print it—regularly, for twelve years."[2]

Ascoli held himself responsible in detail for the contents of the magazine, especially the front of the book. (He normally delegated the back of the book—essays, stories, verse, and reviews—to a senior editor and finally would read it only in page proof.) Occasionally he needed to apologize for mistakes. When the cover of the issue for January 10, 1957, which was supposed to bear the words "Our Gentle Diplomacy," bore instead—through an inexplicable error at the Dayton printing plant—the words "Our Gentile Diplomacy," Ascoli sent letters of explanation and apology to *The Reporter*'s 100,000 subscribers. (It was *not,* he insisted, a Freudian slip.) When the translator of Abram Tertz's *Makepeace Experiment* protested against a "factually incorrect as well as defamatory" passage about her in a *Reporter* review of the book, and the reviewer made a half-hearted apology, Ascoli wrote that in permitting the publication of the passage in question "I have failed to live up to my editorial responsibility."[3]

Ascoli did not originally envision himself as the editor of *The Reporter.* Describing in 1971 the genesis of the magazine, he said:

> Around 1945 or 1946 I got tired of teaching. I felt there was a gramophonical quality to teaching. I had written about politics, and I had the idea that there was a need for a kind of political reporting, political analysis, that was different from the traditional type of magazines of opinion. I was trying to see how a liberal approach—a liberal political philosophy applied to current events, to public events, to squeeze the *meaning* of current events—could be developed. . . . I went from one failure to another. I was afraid of publishing—I hadn't the remotest intention of doing it by myself.

Ascoli's first attempt was a consultation with one of his associates from *Free World* magazine and with two or three wealthy New Yorkers who were interested in backing a new magazine. Then at the end of 1946 one member of the group, Michael Straight, unexpectedly hired Henry Wallace as coeditor of the *New Republic.* "I have always been an anticommunist," said Ascoli, "and I did not see how I could remain in a group where one of the members was coeditor of another publication and therefore boss of Henry Wallace; so I left."[4]

Some of this group were subsequently involved in the founding of *United Nations World,* a new monthly resulting from a somewhat uneasy merger of three magazines: *Free World, Asia,* and *Inter-American.*[5] After six and a half years of publication *United Nations World* went out of existence, to be succeeded the following month by *World* magazine. *World* lasted only from November 1953 to June 1954, and the month before it folded Ascoli announced that the subscription list of *World* had been taken over by *The Reporter.*[6]

After he had left the *United Nations World* group, Ascoli and his wife decided to finance their new magazine themselves and established a limited corporation with a capital of $1.5 million.[7] The next problem for Ascoli was finding an editor: "I was thinking always that to be an editor was sort of an act of arrogance. I could not be the editor. I had been a professor. What did I know about editing? I was thinking of myself as a contributor, a writer, and nothing more." He talked about his projected magazine with several friends, and one of them, James Reston, suggested as editor Wallace Carroll, a former European correspondent for the United Press who had been with the Office of War Information during World War II. Ascoli then named Carroll to head a skeleton staff for an unnamed magazine in Washington, D.C. Ruth Ames, the magazine's extremely capable librarian, who joined the staff in October 1947 and remained on Ascoli's staff after the magazine folded, has described its prepublication days:

> There was something known loosely as an editorial board, which met irregularly. It included James Reston, Eric Sevareid, and infrequently, Kenneth Galbraith, Arthur Schlesinger, and Frank Jamieson, public relations man for Nelson Rockefeller, when brought by Ascoli. For editorial conferences we all sat around a big table, talking generally about the magazine, and specifically about news events, how they should be treated, what articles should be in the works, and so on. There was a lot of talk, and for one recently from Ogden, Utah, it was pretty heady stuff.
>
> There was much discussion about a name for the magazine. Wally at one time had a staff contest, but none of the entries

> seemed right. You had to be careful not only of copyright problems but of associations with other magazines, living and dead. Finally Dr. Ascoli thought of what I have always considered the perfect name: THE REPORTER.[8]

Ascoli made nearly weekly trips to Washington for consultation, but after he broke his hip in May 1948, the trips had to be discontinued for a period. In the fall of that year he decided to take the magazine to New York, but Carroll was not willing to leave Washington.[9] So on the advice of Galbraith and Schlesinger—"One evening in my house they told me, 'What the hell are you doing, looking like a desperate man for an editor?'" —Ascoli decided to edit the magazine himself.

The first issue of *The Reporter,* dated April 26, 1949, appeared on April 19 of that year. Some 50,000 copies went for sale on newsstands, and 125,000 free copies were mailed to a list of educators, businessmen, labor leaders, and government officials.[10] The first issue contained a five-page prospectus, written by Ascoli, that set forth the aims of the magazine:

> We are publishing *The Reporter* because we think there is room in the United States for a type of reporting free from obsession with headline "news" and from the conceit of "opinions." . . . *The Reporter* is to be a magazine of facts and ideas, not of news or of opinion. . . . We shall deal with situations, but not in order to make ourselves the mouthpieces of the nothing-can-be-done-about-it school of thought. Above all, we shall deal with policies, and our reporting of facts and ideas will be aimed at promoting the constant development of an American policy adequate to the responsibilities and to the limitations of America in the present moment of history. . . . It is a magazine for the citizen and not just the reader. . . . Of our readers we ask that they consider themselves members of our group, for we think of them as partners, not customers to be courted.

Published every two weeks, *The Reporter* was initially described on its masthead as "A Fortnightly of Facts and Ideas," and after 1955 as "The Magazine of Facts and Ideas." Ascoli liked Alfred North Whitehead's phrase, "ideas about facts."[11] Facts "are not born free and equal," Ascoli has insisted. "They have to be examined," and the "criteria for selecting, classifying, and testing facts are called ideas." Without facts, ideas are empty abstractions; without ideas, facts are "unorganized tidbits." The two must interact if one is to understand the meaning of events.[12] When some critics complained that the magazine was not sufficiently "topical," Ascoli responded, "*The Reporter* does not get out of breath competing with newspaper headlines, and is not content with rewriting them."[13] A few years later he wrote that *The Reporter* liked "to concentrate on what really matters"; its readers knew "that we try to cover everything that is important—no matter whether it makes news or not."[14]

The first issues stated that no advertising would be carried for the first six months until it was clear what sort of audience the magazine had. Questionnaires about the magazine were provided for readers to fill out and send in. To stimulate a dialogue with readers, questions were posed, such as "What do you think can be done to improve the American public's knowledge of foreign affairs?" Prizes of $25 were given for the best letters in response, and the letters were printed in subsequent issues.

To its early critics, the aims of *The Reporter* seemed impossibly high, and it appeared too earnest in trying to realize them. *Time* complained that nearly half of the thirty-six pages in the first issue were devoted to "a leisurely, dull analysis of President Truman's program for developing the world's backward lands."[15] One early reader wrote: "It would . . . be beneficial if you relaxed your style at times and did not take everything so damned seriously."[16] During the first months each issue had a theme, such as labor, the Cold War, or civil liberties. The issue that the staff came to remember most vividly was one devoted to agriculture (July 19, 1949). "Max used to joke about it," recalls one former staff member. "It was the most boring, unreadable thing. Parities. What the hell are parities, you know?"

The first issue of *The Reporter* listed on its masthead an editorial staff of about forty, a majority of them identified as research writers or staff assistants. Claire Neikind, one of the seven original senior writers, had written for the *New Republic* in 1947 and 1948. In the early 1950s, after her marriage to the writer Tom Sterling, she moved permanently to Italy and covered Europe, the Middle East, and Africa for *The Reporter* throughout the life of the magazine. Barbara Carter, one of the original research writers—Ascoli considered her "superb" at checking facts—began getting bylines in 1961 and continued to contribute after she left the staff in the mid-1960s. Robert Bingham, another research writer, had graduated from Harvard in 1948 and had been briefly and unhappily on the staff of *Time.* After writing a few early political articles, he began to work with a senior writer, Gouverneur Paulding, on the back of the book. In May 1957, Bingham succeeded Al Newman as managing editor and kept this position until he left at the end of 1964 to become an editor of nonfiction pieces at the *New Yorker.*[17] There, as an illustration of what Bingham calls "the curious ties between *The Reporter* and the *New Yorker,*" he came to occupy the office that had once belonged to Robert S. Gerdy, who had gone from the *New Yorker* to become *The Reporter*'s first assistant managing editor, remained as its managing editor from 1950 to 1952, and returned to the *New Yorker* in 1953.[18]

Douglass Cater, a friend of Bingham's from Harvard, had preceded Bingham to *The Reporter* and later became one of its best-known staff members, with a reputation as "one of the best of the Washington correspondents."[19] A veteran of the Office of Strategic Services and one of the founders of the National Student Association in 1947, Cater soon was given the position of Washington editor, a position he kept—with some time off for teaching at Wesleyan University and for writing two influential books, *The Fourth Branch of Government* (1959) and *Power in Washington* (1965)—until he left to become special assistant to President Johnson in 1964.[20] At that time he was, according to a White House colleague, "one of the few figures of the American intellectual or quasi-intellectual world who seemed ready to become a thoroughgoing LBJ man."[21] Ascoli said in 1971: "The Washington bureau operated as an entity dependent directly on me and in constant communication with me. We had an hour or two conversation every Sunday."

Because many of the original staff were young, and because many of the articles were collective efforts, the original policy of *The Reporter* was to print staff-written articles anonymously and to give bylines only to outside contributors. The only two bylined articles in the first issue were by August Heckscher and Alfred Kazin. Readers were quick to protest against this anonymity—some said that it led to "arrogance"—and by the tenth issue (August 30, 1949), all staff-written articles (except an occasional collaborative piece) bore their writer's initials, and within a few months all articles bore regular bylines. Outside contributors in the early months included some well-known names—A. A. Berle, McGeorge Bundy, James MacGregor Burns, Theodore Draper, Ralph Ellison, Herbert Elliston, John Kenneth Galbraith, A. Whitney Griswold, Joseph C. Harsch, Max

Lerner, Saul Padover, James Reston, Arthur Schlesinger, Jr., Ignazio Silone, Wallace Stegner, and Harold Taylor.[22] And if some of the thematic issues were uninspiring, others, such as "Politics vs. the People," "The UN—Policies and Personalities," and "The Negro Citizen," showed imagination.

Still, as one former staff member puts it, "it was very heavy going until about the time of the China Lobby thing." The China Lobby story, a long two-part investigative article in the issues of April 15 and April 29, 1952, drew widespread attention to *The Reporter* and led to a dramatic rise in its circulation. It was in large part the product of a staff member, Philip Horton, who seems to have been almost indispensable to the success of *The Reporter.* Horton, who was brought to the magazine as managing editor when it was about five months old, bore several titles before being named executive editor in the spring of 1958. But he seems to have served virtually as Ascoli's second-in-command for nearly the whole duration of the magazine, even though Ascoli brought in three editors from the outside at various times to serve as Horton's nominal superiors—Harlan Cleveland as executive editor for two years from March 1953 to June 1955,[23] Irving Kristol as editor for one year from late 1958 to late 1959, and Dwight Martin as coeditor for a few months from September 1963 to May 1964. Horton is a very different kind of man from Ascoli—they severed relations in the final months of the magazine, and one suspects that they never were very close personally—but one former staff member who knows both men well maintains that Ascoli's concept of the magazine could never have become a reality without Horton. Nevertheless, Ascoli was disappointed that he seemed increasingly unable to have any serious discussion of ideas, political or otherwise, with Horton as he could with other staff members such as Gouverneur Paulding, William Lee Miller, Douglass Cater, and Harlan Cleveland.

Horton, a Princeton graduate who had joined the Harvard faculty as English instructor in the late 1930s, had gained some recognition in the literary world as author of a pioneering biography of Hart Crane, published in 1937, only five years after the poet's death. During World War II Horton served with the Office of Strategic Services in London and Paris. After the war he remained briefly in Paris as chief of the Central Intelligence Agency liaison mission with the French intelligence service. Upon his return from Europe he spent a year as associate editor in the foreign news section of *Time.* Three months after he joined

Time Horton developed serious ulcers—"I got rid of them without a trace two weeks after I left; it was a very interesting psychosomatic demonstration"—and he began looking around for something else.[24] Two mutual friends of Ascoli and Horton—Arthur Schlesinger, Jr., whom Horton had known at Harvard and who had served as Horton's deputy in the OSS in Paris, and August Heckscher of the *New York Herald Tribune*—tried to interest him in working for *The Reporter.*

At first Horton, who had looked at some of the early issues of the magazine—"I remember one was on some incredible thing like farm policy" —was not interested. He considered the thematic approach a disaster. But his friends persisted, and

> I was getting sort of curious about this man Ascoli. . . . Well, after one or two meetings I could see that he had in mind a quite different sort of magazine—one that I was quite interested in, because during the war years and after the war in Paris I was very much struck by the gap in the journalistic spectrum, and I realized that there was really nothing in between the *Nation* and *New Republic* over on the left and the news magazines—*Time* and *Newsweek*—I won't say on the right, but on the conservative side. . . . And in the talks I then began having with Ascoli I could see that that was what he really wanted [to fill the gap], and our political philosophies at that time were quite compatible. He felt just as strongly about the doctrinaire liberal as I did—perhaps even more so.

After Horton joined *The Reporter*, he and Ascoli divided responsibilities in a way that, in Horton's words,

> worked out beautifully as far as I was concerned. I had primary responsibility for planning and lining up copy—choices of subject and all the rest of it—for oncoming issues. Ascoli concentrated very heavily—almost exclusively sometimes—on what we called "A" issue, namely, the issue going to press. This left me complete freedom to operate as I wanted to operate: to follow my hunches and leads and recruit the writers. I tried over the years to put together a stable of really good correspondents, whether free-lance or staff, both abroad and

in this country. So I was able to line up people like Ed Taylor, who had been an old personal friend, and Denis Warner in Southeast Asia.

Ascoli's account of their working relationship differs slightly in emphasis:

> I concentrated on writing my editorials, most of the Notes, cultivating my own writers, and doing the overall editing job, while Phil pursued his investigative hunting. But in what I had to edit, some of the hardest parts were the literary result of his investigations. Each one had needed a writer, and a good one. . . . In general, relations with our foreign correspondents as well as with the Washington bureau were almost entirely in my hands—the "almost" due to Denis Warner, whom I came to know only during the last year. . . . As for the other exposé stories after the China Lobby, Horton's investigative role was lessened because of the work done by a technician in the field who had worked for the FCC, Charles Clift. . . . In the meantime, Horton was going around meeting all sorts of people, as an intelligence officer must.[25]

George Bailey, whose coverage of Germany in *The Reporter* was to win an Overseas Press Club Award in 1960, had never done any journalistic writing when he first approached the magazine in 1956. He had been turned down by a succession of New York editors, but Horton was interested and told him to start sending in copy. Bailey gives Horton credit for teaching him how to write.

Ascoli pays tribute to Horton's immense industry in getting contributors—even if "sometimes I couldn't accept what they were doing" —and to one of his personal qualities which Ascoli recognizes he himself does not possess: "Phil has the merit of cultivating up to a point a story like the China Lobby with an endless patience." As Horton describes the handling of the China Lobby story:

> I handled that all by myself for almost a full year, before assigning the story to a writer, corresponding with people all over the world. And I ended up with three fat file folders. It was a major exposé of something people talked about for

> ages. I got a little bit irritated seeing this constant reference to the China Lobby, and obviously the natural question finally occurred to me: "If there is one, why the hell hasn't it been written up?" I had a good many friends in the State Department in those days . . . and because of the McCarthy thing I had been in close touch with them, particularly with the Far Eastern ones. So it was no trouble at all to start checking with them. I always described it as a very long-drawn-out piece of mosaic work.

Ascoli comments:

> Actually, the idea of a China Lobby story had first occurred to me, but that means nothing compared to the enormous amount of work Phil did. For a year, it was sheer joy to hear of his meetings with the most improbable assortment of characters, Chinese and Americans. . . . He learned to pronounce Chiang in a way that he said was pure Chinese.[26]

The assignment was then given to Charles Wertenbaker, who wrote the bulk of it, and Ascoli did the final editing, working into the early morning hours at Dunellen, New Jersey, where *The Reporter* was then printed.[27]

Two years before *The Reporter*'s exposé, a writer in the *New Republic* had called for "a Congressional inquiry into the irresponsible and dangerous activities of the pro-Nationalist China Lobby,"[28] and Senator Wayne Morse in the summer of 1951 had called for an investigation, which was subsequently conducted quietly by the Departments of State, Treasury, and Agriculture.[29] But this quiet investigation was moved dramatically into the public domain by *The Reporter.* When the first installment appeared—in an issue with a bold red dragon on the cover—Horton got a call from a Washington newsman who said, "You ought to know that the Chinese Embassy has people out buying up these issues and they're burning them in trash baskets along Connecticut Avenue." This free publicity, of course, was invaluable to the magazine.

Horton seems to have had little taste for some of the more onerous tasks of editing and little interest in writing, although he occasionally produced a good investigative article, such as his story of United States policy toward Yemen, "Pursuit of a Mirage" (October 24, 1963). His major contributions to *The Reporter* stemmed from the fact that he was, in Ascoli's words, "an extraordinarily industrious intelligence officer." He deserves credit for inspiring several scoops.

He assigned staff members William Fairfield and Charles Clift to write the series "The Wiretappers," which won *The Reporter* its first major award, a special George Polk Magazine Award, in 1953. "Then I assigned Dwight Macdonald what I think was the first article on the lie detector in government. I had quite a time persuading Dwight that it was really an important story." The resulting two-part article, "The Lie Detector Era" (June 8 and June 22, 1954), won the Benjamin Franklin Magazine Award in 1955. An article by Robert Bendiner, "The Engineering of Consent—a Case Study," which won a Benjamin Franklin Magazine Award the following year, had also been assigned by Horton. What one staff member considers Horton's biggest scoop was an article he assigned to Noah Gordon, science editor of the *Boston Herald*, on Timothy Leary and his LSD experiments at Harvard. The story, "The Hallucinogenic Drug Cult" (August 15, 1963), which may have had a part in Leary's subsequent dismissal from Harvard, drew several columns of appreciative letters from *The Reporter*'s readers.

A third man who played a major role in shaping *The Reporter* was Gouverneur Paulding, one of the original staff members, who, under a variety of titles—senior writer, associate editor, senior editor—bore much of the responsibility for the back of the book. A great-great-grandson of the nineteenth-century American writer James Kirke Paulding, he was a Harvard graduate who had stayed on in Europe after serving in World War I. During the 1930s he served for several years on the staff of Emmanuel Mounier's Personalist journal *Esprit*. Equally at home in French and English, he translated extracts from the writings of the Swiss novelist and essayist, C. F. Ramuz, under the title *What Is Man* (1948), and with his wife, Virgilia Peterson, translated Father R. L. Bruckberger's *Image of America* (1959). A devout Catholic, he had been managing editor of *Commonweal* during the mid-1940s—the period when Ascoli was contributing several articles to that magazine. In the early issues of *The Reporter* he wrote a column entitled "To Man's Measure" and later handled many of the book reviews.

Paulding was particularly close to Ascoli personally, and until his illness—he died of leukemia in 1965—he was the only person Ascoli allowed to assist him in formulating the ideas and phrasing of his editorials, which Ascoli sometimes found it agony to compose. Ascoli may have felt a philosophical rapport with Paulding that he did not always feel with others on the staff, for Paulding seems to have been close to the kind of European liberalism with which Ascoli identified himself. A liberal, Paulding once wrote, "knows that he is soft and that he has not the Communist sense of purpose." Yet he is constantly "at work breaking down the rigid forms established by the cynical, the lazy, the experienced, and the fatigued. . . . Without him the nation accepts the absolutes of right and left, the easy answers."[30] Summarizing the message of the writings by Ramuz that he had translated, Paulding described a philosophy that one suspects was close to his own:

> It is that first things come first. The first things are not in the realm of economics, or politics, or social service, or ethics, or war. The first things are not in the theory and practice of Communism, or democracy, or distributionism, or "back to the land." They do not have any connection with the possession of the atom bomb; or the gold in Fort Knox. They do not regulate production, or advertising, or public relations. The first things are things of the gods or of God. . . . It is not until these primary matters are settled that one can proceed with a civilization.[31]

Paulding's personality—which seems to have included a sweetness and gentleness that neither Ascoli nor Horton could be accused of possessing in conspicuous measure—may have been as important to the magazine as his ideas. Former members of the staff remember him with special warmth and affection. After his death Ascoli wrote: "He had been with this magazine since the beginning, already a close friend of mine when he joined the staff. . . . Well before his illness, our care for Gov, and of Gov for all of us, was one of the strongest elements in keeping our group together."[32]

Somehow these three quite different men, in combination, helped create an atmosphere in which the magazine could flourish. "In 1955, when I came to the magazine," says one former staff member, "the editorial staff was completely devoted to the magazine. It was the most unified group of people I had ever met." This unity is the more striking because of the diversity and strong personalities of some of the prominent staff members.

Marya Mannes, who joined the staff in 1953 and stayed for nearly ten years, brightened the magazine considerably through her regular column of television criticism, by a variety of articles which she once described as "riding herd on sacred cows," and perhaps most of all by the topical satirical verse she contributed pseudonymously as "SEC." In 1959 she received the George Polk Memorial Award for her television criticism in *The Reporter,* and the same year the magazine received a National Brotherhood Certificate from the National Conference of Christians and Jews for her article "The Murder of In Ho Oh" (January 26, 1958), a compassionate account of the senseless murder of a Korean student by a juvenile gang in Philadelphia. As she described herself in 1958, "I fail to fit neatly into any social or artistic categories. Conservatives will undoubtedly shout 'liberal!' but the more militant liberals in both art and politics will in the same breath cry 'reactionary!' . . . The big magazines find me too special and controversial."[33] She and Ascoli had a sharp falling-out over Vietnam in the summer of 1963. As a result, in her words: "I left, or was fired from, *The Reporter.* I could no longer hide my deep political differences from my editor."[34] But they were soon personally reconciled, and she and Gouverneur Paulding were the two former staff members to whom Ascoli paid tribute by name in his final editorial.

Robert Bendiner, who had been managing editor of the *Nation* from 1937 to 1944 and its associate editor for several years after World War II, and who had been one of the founders of the ADA, contributed articles to *The Reporter* from 1951 through 1965 and was listed as contributing editor on the masthead from 1956 through 1960. Many of his forty articles were concerned with politics or political campaigns. Two of his longer articles were his award-winning "The Engineering of Consent" and his study of automation, "The Age of the Thinking Robot, and What It Will Mean to Us" (April 7, 1955). In the mid-1960s he joined the editorial board of the *New York Times.*

William Harlan Hale, who was listed as contributing editor from late in 1955 to early in 1958, had been briefly a member of the original staff, having come from the *New Republic,* where he had been a senior editor. In an article for that magazine Hale had tried to resolve the conflict between the "economic" and "noneconomic" strains in American liberalism:

> We have had in our ranks a cult of freedom and a cult of control. Perhaps it is time that we went back to the first principle which

> underlies both our ideas of freedom and our idea of control—the principle that for all his acts, a man is to be held strictly accountable to his fellow man. . . .
>
> How these ideas can effectively be put into practice under any system less rigorous than socialism, this particular deponent does not see. But although socialism has often been called "inevitable," without a moral basis I can't see a reason for wanting it.
>
> To work out this moral basis may yet become the function of those whom we call liberals.[35]

After serving in Austria from 1950 to 1953 as public-affairs officer for the State Department, Hale contributed nearly forty articles to *The Reporter* until 1958, when he became managing editor of the new *Horizon* magazine. A number of his articles dealt with the State Department, including a critical examination of John Foster Dulles, "The Loneliest Man in Washington" (October 18, 1956) and of the State Department security chief under Dulles, Scott McLeod, " 'Big Brother' in Foggy Bottom" (August 17, 1954). His *Reporter* article "Going Down This Street, Lord" won a National Brotherhood Certificate from the National Conference of Christians and Jews in 1956.

Theodore H. White—whose career has included a stint in China with the Luce publications, six months with the *New Republic*, and authorship of *The Making of the President—1960* and its quadrennial successors—began writing for *The Reporter* in the fall of 1950 while he was with the Overseas News Agency in Europe. (One of his earliest pieces—in the issue for November 21, 1950—was "Indo-China: A Tiger by the Tail.") By the fall of 1951 he was described to readers as "a regular European correspondent for *The Reporter.*" In March 1954, he was first listed on the masthead as national correspondent, and one of his major articles, "Texas: Land of Wealth and Fear" (May 25 and June 8, 1954), won a Benjamin Franklin Magazine Citation in 1955. The last of his thirty-five articles for the magazine appeared in February 1955, and two months later his name disappeared from the masthead.

William Lee Miller, who taught Christian ethics for several years at Yale and who had been involved in local politics in New Haven before moving to Indiana University as professor of religious studies and political science, contributed some forty articles between 1953 and 1967. He was listed as staff writer in 1955 and 1956 and again in 1958 and 1959. "Although Mr. Miller is well-known among the younger American theologians, it is in our pages that he has reached a broader audience for the first time," *The Reporter* commented in 1953. "He is the kind of young man we are constantly looking for."[36]

Paul Jacobs, one of the more radical *Reporter* staff members, began contributing articles on civil liberties—"Should Communists Be Allowed to Eat?" (March 24, 1955) and "The Nightmare Adventure of Irving Markheim" (November 17, 1955)—and was first listed as staff writer in January 1957, when his two-part article "The World of Jimmy Hoffa" appeared. The nearly thirty articles he contributed before he left the staff in 1961 dealt largely with labor, civil liberties, and civil rights, but his most influential article was "Clouds from Nevada" (May 16, 1957), an exposé of the radioactive fallout from nuclear testing. It was cited in major stories by the *Washington Post,* the *New York Post,* and the *Daily Worker*; it drew an appreciative letter to *The Reporter* from Adlai Stevenson; and it won a Sigma Delta Chi Award in 1958 for distinguished public service.

The gradual disappearance of these writers from *The Reporter* after 1960 coincided with the increased prominence in its pages of Meg Greenfield, a gifted political writer and superb satirist who succeeded Douglass Cater as the magazine's Washington editor in the spring of 1964. A 1952 graduate of Smith, she began her career at *The Reporter* as a "mousy" member of the library staff. ("She seemed mousy," says one former staff member. "My God, were we wrong!") Several months after the appearance of her first major article, "The Prose of Richard Nixon" (September 29, 1960), she was listed on the masthead as staff writer, and as one of *The Reporter*'s star attractions soon had the privilege of making last-minute revisions in her articles "in a way," says one former managing editor, "that infuriated those of us who had to handle the mundane aspects of journalism." But, says another staff member, "it was usually something very, very good." Her political position was apparently close to that of Ascoli, and after *The Reporter* folded she went with him briefly to *Harper's* before going to the *Washington Post* and *Newsweek.*

Another staff member who gained increasing prominence in the 1960s was George Bailey, who had been a correspondent for *The Reporter* from Germany and Eastern Europe since 1957. After June 1961, he was listed as staff writer, and after Horton left the magazine in April 1968, Bailey was listed as executive editor for the magazine's final three months. Like two others of *The Reporter*'s four chief foreign correspondents, Claire Sterling and Edmond Taylor, Bailey was an American expatriate who had spent much of his adult life in Europe. (The fourth, Denis Warner, who began contributing in 1957 and became the regular correspondent from Southeast Asia, is Australian.) A proficient linguist who had studied under C. S. Lewis at Oxford, Bailey had served as General Walter Bedell Smith's interpreter in Russian and German at the surrender negotiations in Rheims in 1945. After several years as liaison officer to Soviet forces in Germany, he served in the Pentagon from 1954 to 1956 as senior consultant on Soviet affairs.

Edmond Taylor, a friend of Horton's who had contributed to *The Reporter* since 1953, was listed on the magazine's masthead as staff writer from November 1959 until *The Reporter*'s demise. A native of St. Louis who had been foreign correspondent for the *Chicago Tribune* and later for CBS in France, Taylor served with the OSS in Southeast Asia during World War II and later briefly with UNESCO and with the U.S. Psychological Strategy Board. After his return from Europe in the early 1950s he contributed to the *Washington Post* and the *New Republic* as well as to *The Reporter,* where more than a hundred of his articles appeared during the fifteen years he wrote for the magazine.

In addition to its staff *The Reporter* had a number of regular outside contributors. Among those who wrote most frequently for the front of the book were William Hessler of the *Cincinnati Enquirer,* who contributed nearly fifty articles between 1950 and 1964; Eric Sevareid, more than sixty of whose CBS radio broadcasts were printed in the magazine from 1954 to 1961; and Isaac Deutscher, the Polish-born expert on Communism, who contributed more than fifty articles between 1950 and 1961. Adolf A. Berle, a close friend of Ascoli's, authored over twenty articles between 1950 and 1965, and Henry Kissinger, another close friend of Ascoli's, contributed eleven pieces between 1957 and 1967.[37]

Frequent contributors to the back of the book included book reviewers John Kenneth Galbraith, who appeared more than fifty times between 1950 and 1966; Alfred Kazin, who appeared in the first issue, fairly regularly from 1958 through 1962, and sporadically thereafter until 1966; Sidney Alexander, who appeared frequently from 1955 to 1961; and George Steiner, who appeared from 1959 through 1965. Drama criticism was contributed by Gerald Weales of the University of Pennsylvania from 1956 through 1968, by Tom Driver of Union Theological Seminary from 1962 through 1964, by Gore Vidal from 1957 to 1965, and others. Art criticism by Hilton Kramer, a former editor of *Arts* magazine, appeared from 1959 to 1965; music criticism was contributed by Roland Gelatt of

High Fidelity magazine (1955-68), Fred Grunfeld (1959-67), and Nat Hentoff (1958-65). Hentoff, who had described himself in a letter to the editor in 1955 as "a subscriber and long-term admirer of the political (though not aesthetic) insights of your magazine,"[38] had been invited by Bingham, on the basis of a subsequent letter to the editor, to write for the magazine. He contributed more than seventy articles, many of them on jazz,[39] before fulminating in the *Village Voice* in 1965 against "General Ascoli" who "has intensified his role as interpreter of real politik to us soft-at-the-center heretics."[40]

In addition to such regulars, *The Reporter* published articles, reviews, stories, or poems by more than seventeen hundred other contributors, a majority of whom appeared fewer than five times in the magazine. Although several hundred of them were well-known figures in the political, journalistic, or academic worlds, a substantial number were young writers being published for the first time. Among the contributors from political life were eleven United States senators—Clifford Case, Frank Church, J. William Fulbright, Hubert Humphrey, Jacob Javits, John Kennedy, Wayne Morse, Richard Neuberger, Hugh Scott, Stuart Symington, and Millard Tydings—and eight representatives: Emanuel Celler, Peter Frelinghuysen, Jr., Harris Ellsworth, Eugene McCarthy, Charles Matthias, Henry Reuss, Stewart Udall, and Harrison Williams. Contributors who were, had been, or were to become prominent members of the executive and judicial branches included Dean Acheson, Chester Bowles, McGeorge Bundy, John Paton Davies, William O. Douglas, Arthur Goldberg, Averill Harriman, George Kennan, Daniel P. Moynihan, Paul Nitze, William Rogers, Eugene and Walt W. Rostow, John Carter Vincent, G. Mennen Williams, and Adam Yarmolinsky.

More than 350 American journalists on the staffs of newspapers, the wire services, radio and television networks, and other periodicals—as well as many free-lancers—contributed to *The Reporter* over the years, including forty from the staff of the *New York Times,* twenty-four from the *New York Herald Tribune,* twenty-two from the *Washington Post,* and thirty from *Time* and *Newsweek.*[41] Nearly a hundred journalists from overseas publications, many of them British, also contributed to *The Reporter,* some of them fairly frequently, like Gordon Brook-Shepherd of the *London Daily Telegraph* (1958-64), Alastair Buchan of the *London Observer* (1955-61), J. H. Huizinga of the *Nieuwe Rotterdamsche Courant* (1950-67), and John Rosselli of the *Manchester Guardian* (1951-67).

The Reporter served as a stepping stone in the careers of a few young journalists. Its first European correspondent (1950-52), whose writing for *Le Monde* had caught the eye of Horton while Horton was still at *Time,* and whom Leland Stowe, *The Reporter*'s first foreign editor, had met in Paris, was Jean-Jacques Servan-Schreiber, who was later to found *L'Express* and to attract widespread attention in the late 1960s as author of *The American Challenge.* A former managing editor of the Harvard *Crimson* who had gone to work for the West Point, Mississippi, *Daily Times Leader* began submitting articles on the Southern racial crisis to *The Reporter* in 1955. Some of them interested James Reston, and he hired their author, David Halberstam, for the *New York Times* Washington bureau.[42] Another Pulitzer Prize-winning journalist, J. Anthony Lukas, also contributed articles to *The Reporter* in the course of his journey from the Harvard *Crimson* to the *New York Times.*

Nearly 250 contributors were college or university faculty members—more than half of them in the social sciences. They included not only well-known academic names like Daniel Aaron, Jacques Barzun, Daniel Bell, Hennig Cohen, Henry Steele Commager, Gordon A. Craig, Marcus Cunliffe, Carl Degler, Benjamin De Mott, F. W. Dupee, Leslie Fiedler, Nathan Glazer, Andrew Greeley, Andrew Hacker, Robert Heilbroner, Richard Hofstadter, Irving Howe, Sydney Hyman, Seymour M. Lipset, Perry Miller, Howard Nemerov, Justin O'Brien, Saul Padover, John Roche, Clinton Rossiter, Paul Samuelson, S. Fred Singer, John William Ward, and Donald S. Zagoria, but also a heartening number of graduate students, instructors, and assistant professors, some of whom were being published for the first time.

The predominance of political, journalistic, and academic figures among the contributors and among the authors of letters to the editor is one measure of the character of *The Reporter*'s audience. But the size of the audience is also important, for *The Reporter* came to have a much wider readership than such traditional liberal magazines as the *Nation*, *New Republic*, *Progressive*, and *New Leader.* The circulation of *The Reporter* increased during its nineteen years by more than 1300 percent—from the 15,000 charter subscribers it had at the end of its first year to the more than 200,000 subscribers it had when it came to an end in June 1968. Its initial growth was slow: by the end of 1951, when it was more than two and a half years old, it had a circulation of only 20,000. "The turning point in *The Reporter*'s career came in the spring of 1952, when we published two articles on the China Lobby; almost overnight our circulation shot up from 20,000 to 55,000."[43] In July 1954, just after taking over the subscription list of the defunct *World* magazine, *The Reporter* announced a circulation of more than 100,000. This period of rapid growth, which coincided with the McCarthy era, was followed by a five-year period of relatively slow growth: by its tenth anniversary in 1959 it had a circulation of about 125,000. It claimed to have a renewal rate of almost 70 percent, "one of the highest in the field."[44] A second spurt in circulation came in 1960, when its circulation rose to an estimated 160,000.[45] By the end of 1965 it had passed the 200,000 mark and remained above that level for its final two and a half years.

At the beginning of 1966 the state with the largest percentage of *The Reporter*'s subscribers—16 percent—was California. New York followed with 14 percent. Together the East and West Coasts accounted for more than half of *The Reporter*'s circulation. About 6 percent of its subscribers lived overseas, and another 6 percent in Virginia, Maryland, and the District of Columbia.[46] In 1961 Ascoli noted proudly that *The Reporter*'s circulation in Washington, D.C., was "higher than that of several of the old, established news or business or anthology magazines." He added, "*The Reporter* will never have what is called a mass audience, for both its readers and writers are men and women who feel responsibility toward the masses, and therefore, when they have to, act according to their conscience."[47]

Some two years earlier the editor of *Harper's* had made a similar statement—although in far different language—about the intended audience for his magazine—"that one percent of the population which might be described as the decision-makers, opinion-formers, and taste-setters . . . the leadership elite group."[48] And just as there was a difference in the language of the editors, so was there a difference in the interests of those attracted to the two magazines. In 1958 more than a thousand American opinion leaders at a White House conference on foreign aid were asked what magazines they read. Sixteen percent read *The Reporter* (fourth in popularity after *Time, U.S. News,* and *Newsweek*). Nine percent read *Harper's.*[49] Perhaps even more significantly, a readership survey in 1955 showed that 58 percent of *Reporter* subscribers did not regularly read *Harper's,* the *Atlantic,* or the *Saturday Review,* and 43 percent did not regularly read *Time, Newsweek,* or *U.S. News.*[50]

Among the paid subscribers to *The Reporter* in 1966 were the President and Vice President, seven members of the Cabinet, and one-third of the Senate. William Rivers, a member of *The Reporter*'s Washington staff briefly in 1959 and 1960, later wrote that it was "a magazine so significant that nearly every important government official either reads it himself or assigns it to an underling."[51] In the spring of 1963, after *The Reporter* had published Henry Kissinger's long, complex essay, "NATO's Nuclear Dilemma," a reporter on the *New York Times* told the magazine's promotion director, Shirley Katzander, "Nobody but *The Reporter* would have the courage to do that to its readers. It's one of the most important things that's been printed in this country at this time. And if only fifty people read it, it will be read by the fifty right ones."[52] This praise is not necessarily inconsistent with a criticism of *The Reporter* published in the same year: "Its heart is so openly on the right side and its payments to contributors so generous that it is somewhat startling that the magazine is not more consistently interesting to read."[53]

Those who wrote to praise the first issue of *The Reporter* in the spring of 1949 included Hubert Humphrey and Nelson Rockefeller, both of whom also sent letters on the tenth anniversary, Jacques Barzun, Henry Seidel Canby, Herbert Elliston of the *Washington Post,* Ben Hibbs of the *Saturday Evening Post,* Estes Kefauver, and Sumner Welles. Adlai Stevenson and Jacob Javits sent congratulations on the fifth, sixth, and tenth anniversaries. Harry Truman sent letters on the fifth and tenth anniversaries. Some of the others whose letters were published on one or more of these anniversaries include Clinton Anderson, William Benton, Clifford Case, Paul Douglas, Mark Ethridge, J. William Fulbright, James Gavin, Averill Harriman, Brooks Hayes, Chet Huntley, Ralph McGill, Wayne Morse, Edward R. Murrow, Reinhold Niebuhr, Walter Reuther, Dean Rusk, Hugh Scott, John Sparkman, Frank Stanton, Stuart Symington, and G. Mennen Williams. These names—most of them would be included on any list of liberals of the 1950s or early 1960s—lend support to Ascoli's statement that "there must be some resemblance . . . some bond of congeniality" among his readers. They also—if one reflects on the divergent views in this group concerning Vietnam in the 1960s—illustrate his subsequent statement: "I hasten to add that I do not assume for a moment that *The Reporter*'s readership is of one mind."[54]

In a survey conducted in the early 1960s among correspondents in Washington, *The Reporter* ranked fourth among the magazines used by these journalists in their work—following *Time, U.S. News,* and *Newsweek,* as it had for the opinion leaders surveyed in 1958, and ranking far ahead of the fifth-place source. When the correspondents were asked to rate magazines of politics and opinion for their fairness and reliability, *The Reporter* was in first place with ninety votes. The *New Republic,* with nineteen votes, was second.[55] During a five-month period in the winter of 1962-63, articles in *The Reporter* served as the basis for at least seventeen articles in foreign newspapers, including the *Times* of London, *Le Monde, Le Figaro, Journal de Geneve, Il Giorno* (Milan), *Hamburger Abendblatt,* and the *Toronto Star.*[56]

Statistics, of course, cannot convey the essence of any magazine. Like its editor, *The Reporter* resists easy classification. It was a liberal magazine—but with a difference. Unlike the *Nation* and *New Republic,* it was printed on slick paper; it had attractively illustrated covers and contained distinctive artwork. It came to resemble the *New Yorker* in appearance more than any other magazine, but without the latter's distinctive foliage of luxurious advertisements. Like the *New Yorker* it carried fiction and verse—often by many of the same writers—and this practice represented another divergence from the *Nation* and *New Republic*:

> Our readers should not be surprised to find a short story in this issue, since our magazine is not only political; it is interested in any meaningful human situation that a writer succeeds in communicating. Reporting, analysis, satirical verse, straight poetry, or fiction—the means do not matter.[57]

> A work of fiction can often tell more about the news of the day than a journalist's report or a sociologist's tract.[58]

Among the better-known contributors to *The Reporter* from the literary world were W. H. Auden, James Baldwin, Saul Bellow, Ray Bradbury, Hortense Calisher, Malcolm Cowley, A. B. Guthrie, Jr., Shirley Jackson, Pamela Hansford Johnson, Mary McCarthy, Bernard Malamud, Bill Mauldin, N. Scott Momaday, Alberto Moravia, V. S. Pritchett, Santha Rama Rau, Delmore Schwartz, Ignazio Silone, C. P. Snow, Wallace Stegner, John Steinbeck, John Wain, and Robert Penn Warren. But probably the most important literary function of *The Reporter* was providing an outlet for the unknown writer. Robert Bingham said in 1971 of his days at *The Reporter*:

> We got a lot of manuscripts. Probably the *New Yorker* gets more, but it was a big activity, getting the manuscripts read and back. And scarcely an issue or a month went by that somebody didn't have a manuscript in his hands that he wanted to push at a story

> conference and get Ascoli to consider that was by somebody nobody had ever heard of before. That was a good thing, I think. And Ascoli was quite receptive.[59]

Lois Phillips Hudson, who went on to publish a book of reminiscences and a novel, appeared in print initially in *The Reporter.* Donald Barthelme appeared in *The Reporter* in 1960, several years before he became a regular contributor to the *New Yorker.* Although Robert Ardrey had been known as a dramatist, novelist, and screenwriter, one of his articles written for *The Reporter* from Africa in the spring of 1955—"A Slight (Archaic) Case of Murder"—caused a flurry in the anthropological world and marked his entry into a new career of anthropological reporting.[60]

Bingham recalls that in addition to himself and Gerdy there were "quite a few people who went back and forth" between *The Reporter* and the *New Yorker.* Reg Massie, who became *The Reporter*'s art editor in May 1950, had earlier done some covers for the *New Yorker.* William Knapp, an assistant managing editor of *The Reporter* in the early 1950s, and Derek Morgan, who had been a copy editor, assistant managing editor, and finally senior editor, are among the former *Reporter* staffers who went to the *New Yorker.*

It would be hard to find two magazine editors more dissimilar in manner than Ascoli and William Shawn of the *New Yorker.* Yet they seem to have shared a common concern about the serious crisis in the world and the part their magazines could play in helping meet that crisis. In his prospectus for *The Reporter* Ascoli wrote:

> There are those who, at the conclusion of an analytical effort, like to sit down and cry; there are those who like to make moralistic soap bubbles on the rim of the precipice; and there are those who once they see the danger of the precipice want to have everything and everybody going down to the bottom. Our attitude, on the contrary, is to look for the men who are responsible for a situation and who have power to do something about it.

Joseph Wechsberg of the *New Yorker* recalls:

> In the late 1950s, Shawn would sometimes talk about the moral and political crises of the Western world, and encouraged me to think of profile subjects, MEN WHO COULD START FROM

> SCRATCH AND BUILD A NEW AND BETTER WORLD, as he once cabled. They would have to be idealistic realists, or realistic idealists, WHO ARE SPRINGING UP ALL OVER THE WORLD TO COPE WITH THE CRISIS WE ARE GOING THROUGH.[61]

It would be a mistake, of course, to emphasize the similarities between the two magazines and to ignore the many differences between them. One way of conveying something of the unique flavor of *The Reporter*—although this approach is misleading in one respect, as we shall see—is to look at the contents of the fifth, tenth, and fifteenth anniversary issues, which, except for the anniversary editorials concerned with the magazine itself, can be considered typical. The fifth anniversary issue (April 27, 1954) opens as usual with the two pages of topical, staff-written "Reporter's Notes" commenting on current issues in the news. A short piece of topical verse by "SEC" (Marya Mannes) is included. Three columns of correspondence from readers follow, sandwiched in among book advertisements. A regular two-column feature normally written by the managing editor, "Who—What—Why" (introduced late in 1952) follows, identifying authors of articles and giving brief résumés of the articles' contents. Then follows the table of contents, which lists first the editorial and then under a topical heading, "Geneva and Points West," lists an article by an assistant professor of political science at Pennsylvania State College on the recognition of Communist China ("Geneva: To Recognize or Not to Recognize"), excerpts from documents by the former Governor of Formosa, K. C. Wu, telling why he broke with Chiang Kai-shek's government ("The K. C. Wu Story"), and an article by Peggy Durdin, "Vietnam Awaits Independence and/or Annihilation."

Under the next heading—a regular one, "At Home and Abroad"—five articles are listed: "Canada's Social-Security System for Children" by Richard L. Neuberger; "The People vs. McCarthy" by Marya Mannes; "The Man Who Shut Down the Port of New York," by Sanford Gottlieb; "Israel's Spiritual Climate" (first part of a two-part article) by Isaac Deutscher; and "Cyprus: Nationalism Through a Looking Glass" by "Ray Alan" (a pseudonym).

The back of the book—under the regular heading, "Views and Reviews" —includes a story of India by Christine Weston, an account of a luncheon with Dylan Thomas by Mary Ellin Barrett, an article by James Hinton on

Igor Stravinsky's opera *The Rake's Progress*, and book reviews by Harlan Cleveland, John Kenneth Galbraith, and Edward Posniak—an economist who had been attacked by Senator McCarthy. Under the heading "Book Notes" are two brief staff-written reviews. Seven of the issue's fifty-six pages are devoted to advertisements, almost all of them for books. The chief exception is a two-thirds page advertisement for the Save the Children Federation.

The tenth anniversary issue (April 30, 1959), in conjunction with Ascoli's editorial, carries complimentary remarks delivered by Senator Fulbright at *The Reporter*'s tenth anniversary dinner. The lead article, "Epidemic on the Highways," is a long analysis of automobile safety that places heavy emphasis on the problems of faulty design. It was written by Daniel P. Moynihan and marks the first of his more than a dozen appearances in the magazine. (Irving Kristol considers his efforts with Bingham in publishing this article to be one of his major achievements during his year with the magazine.)[62] It is followed by Peregrine Worsthorne's "Britain's Lonely Road Toward the Summit," Peter Braestrup's "Limited Wars and the Lessons of Lebanon," and a "Reporter Essay" (these were infrequent) entitled "Foreign Policy on Capitol Hill," written by Thomas L. Hughes, administrative assistant to Chester Bowles.

The "Views and Reviews" section in the back of the book includes a short record review, a movie review, a review of the London theater, an essay on the press, and book reviews by Alastair Buchan, Gore Vidal, Alfred Kazin, Edward T. Chase, Gerald Weales, and Sidney Alexander, followed again by the anonymous "Book Notes." The five pages of advertisements include a two-page Marlboro book ad, a one-page Irish whiskey ad, a one-third page automobile ad (Citröen), and a two-thirds page ad for the Foster Parents Plan, Inc.[63]

The fifteenth anniversary issue (May 7, 1964) contains in its correspondence columns three letters commenting on a recent attack by Ascoli on Senator Fulbright—the man who had paid eloquent tribute to *The Reporter* just five years earlier. The lead article is a report from Prague by George Bailey, "Kafka's Dream Comes True." The section headed "At Home and Abroad" includes five articles: "California Democrats: Battle Royal for the Senate" by Bill Stout; "California Republicans: Are the Birchers Taking Over?" by Jim Wood; "Kenya Discovers the Perils of Uruhu" by Claire Sterling; "The Boy Wonder of Illinois Politics" (Charles Percy) by Hal Higdon; and "The Jalopy Nomads" (an article on "the new Okies") by Barbara Carter.

"Views and Reviews" begins with a three-page poem about Africa, "The God Who Eats Corn" by Irish poet Richard Murphy, and includes book reviews by George Steiner, Sidney Alexander, and Robert P. Knapp, Jr., and an article on Duke Ellington by Nat Hentoff. Two new regular features are the "Reporter Puzzle"—a combination of crossword and acrostic invented by two college professors and carried by the magazine since early in 1960—and a cartoon by Fernando Krahn, a Chilean artist. Book advertisements fill a total of about six pages.

In 1956 Ascoli published a selection of articles in the paperback *Reporter Reader* (Doubleday) and in 1960 a much fuller selection in the hardback *Our Times* (Farrar, Straus, and Cudahy). The first anthology is somewhat more topical, and only six of its selections were reprinted in the second: William Lee Miller's "Some Negative Thinking about Norman Vincent Peale"; Ascoli's review of *Witness*, "Lives and Deaths of Whittaker Chambers"; "Elizam—a Reminiscence of Childhood in Ceylon" by Indian poet T. Tambimuttu; Santha Rama Rau's "The Trial of Jomo Kenyatta"; Gouverneur Paulding's review of *My Host the World*, "A Note on Santayana"; and Ascoli's sixth-anniversary editorial, "This Liberal Magazine." The five hundred pages of *Our Times* contain many of the better-known *Reporter* articles of the 1950s, beginning with "The China Lobby." The book won from Louis M. Lyons the accolade: "Altogether, it can be said of these articles that only a few of the rarer *New Yorker* pieces by Berton Roueché and Christopher Rand surpass them in journalistic effect and literary distinction."[64]

The examination both of the five-year anniversary issue and of the contents of *Our Times* misrepresents *The Reporter* in one important way: it does not do justice to the magazine's editorials, which, with the exception of a handful written by Harlan Cleveland or Philip Horton, and one each by Gouverneur Paulding and Douglass Cater, were always written by Ascoli. The editorial stood at the front of the book, just before the first bylined article, and since it was often written quite late and could run one, two, or occasionally three pages—one-page editorials were the most common—the managing editor often had last-minute headaches with the makeup. A few issues carried no editorial, and sometimes the editorial appeared as a signed piece in "Reporter's Notes."

Looking back over the magazine's first decade, Ascoli wrote in 1959, "In our editorial writing, as well as in our reporting, we have taken a stand on such issues as McCarthyism, the China Lobby, atomic warfare, peace-mongering at the time of Suez, or prospective 'peace for our time' at the expense of Berlin and the West."[65] The first three issues are well represented by selections in *Our Times.* The last two, which are most immediately concerned with American foreign policy, are not represented at all. Thus, although Ascoli repeatedly expressed the central concern of *The Reporter* as being America's relations with the rest of the world, the articles chosen for *Our Times* deal predominantly with domestic issues. And although *The Reporter* received various awards from the Overseas Press Club for its coverage of the Middle East (1957), Germany (1960), Latin America (1961), and Southeast Asia (1965), and for best reporting from abroad and best interpretation of foreign affairs (1966), more than two-thirds of the thirty-two awards it received over the years were for reporting on such domestic issues as civil liberties, civil rights, nuclear testing, the inner workings of politics, juvenile delinquency, education, and medicine.

There almost seems to be a disparity between the articles on domestic questions that gave *The Reporter* its reputation and public image in the 1950s and early 1960s and its editor's continuing concern in his editorials with American relations with the rest of the world. (Horton believes the real strength of *The Reporter* to have been its domestic coverage; Bailey, its foreign coverage. Ascoli commented in 1971 on these differing views of his two former staff members: "I believe everybody is a bit egocentric. The difference between foreign and domestic never made much sense to me.") In this apparent tension between its treatment of America's domestic problems and its concern with America's international responsibilities seems to lie the key to *The Reporter*'s relationship with the liberal community. The magazine's rapport with the liberal community on both domestic and foreign-policy issues in the fifties, and its estrangement from important elements of that community—largely over foreign policy—in the sixties, will be the subject of the next two chapters.

NOTES

1. Jay Jacobs, "Death of the 'Reporter,' " *New Statesman*, September 27, 1968.

2. Marya Mannes, *Out of My Time* (Garden City, N. Y.: Doubleday, 1971), p. 208.

3. *The Reporter*, November 18, 1965, p. 6.

4. Cf. the account of "The Great Wallace Debacle" by Bruce Bliven, long-time editor of the *New Republic* and later an occasional contributor to *The Reporter,*

in his autobiography, *Five Million Words Later* (New York: John Day, 1970), pp. 265-72.

5. See "How the *United Nations World* Came About," *United Nations World*, February 1946, p. 12.

6. *New York Times*, May 22, 1954, p. 18.

7. *Current Biography*, February 1954, p. 9. Marion Ascoli, before their marriage, had been one of the most generous financial backers of another kind of experiment in liberal journalism, the newspaper *PM.* See Kenneth Stewart and John Tebbel, *Makers of Modern Journalism* (New York: Prentice-Hall, 1952), p. 416.

8. Letter from Ruth Ames to the author, March 16, 1971.

9. Carroll went to the *Winston-Salem Journal* as executive news editor for six years, spent from 1955 to 1963 as member of the *New York Times*'s Washington bureau, and returned to the *Winston-Salem Journal* as its editor and publisher. He contributed some satirical verse to *The Reporter* in 1964 and was one of the many who wrote to Ascoli after the closing of *The Reporter* was announced.

10. *Current Biography*, February 1954, p. 9.

11. Ascoli, "A Great Book," *The Reporter*, November 4, 1954, p. 41.

12. Ascoli, "The Hidden America," *The Reporter*, December 1, 1955, pp. 16-17, and interview, January 13, 1971.

13. *The Reporter*, August 2, 1949, p. 40.

14. Ascoli, "*The Reporter* in 1953," *The Reporter*, December 23, 1952, p. 7.

15. "Cub Reporter," *Time*, April 25, 1949, pp. 53-54.

16. *The Reporter*, July 19, 1949, p. 40.

17. Newman had been a war correspondent for *Newsweek* and a copy editor on the original staff of *The Reporter.* Bingham's successor was Donald A. Allan, who had been a senior editor at *Coronet* and a European correspondent for *Newsweek* and the United Press. Allan was managing editor from 1965 until *The Reporter*'s end.

18. Gerdy, a disciple and friend of Thomas Merton, was highly regarded by his colleagues on both magazines. Ascoli says that his own editing load increased after Gerdy left the staff. When Gerdy died of a heart attack at the end of 1965, his colleagues at the *New Yorker* paid an extraordinarily warm tribute to his abilities and character. (Letters to the author from Ruth Ames, March 16, 1971, and Max Ascoli, April 23, 1971; *New Yorker*, January 8, 1966, p. 112.)

19. Edwin Emery, *The Press and America* (Englewood Cliffs, N. J.: Prentice-Hall, 1962), p. 642.

20. *The Reporter*, August 30, 1949, pp. 15-16, and May 21, 1964, p. 12. *New York Times*, May 22, 1964, p. 5.

21. Eric Goldman, *The Tragedy of Lyndon Johnson* (New York: Knopf, 1969), p. 267. For Cater's subsequent career at the Aspen Institute, see Peter Schrag, "Douglass Cater's Secret Mission," *MORE*, June 1975, pp. 10-14, 25.

22. One of them, Theodore Draper, appeared briefly on the masthead as associate editor at the end of the 1950s; most of his articles had appeared prior to that time.

23. Cleveland was named publisher of *The Reporter* in June 1955, the only person other than Ascoli to bear that title. A year later he left to become dean of

the Maxwell Graduate School of Citizenship and World Affairs at Syracuse University but still contributed an occasional article.

24. This and subsequent quotations from Horton are from the author's interview with him in New York in January 1971.

25. Letter from Ascoli to the author, April 23, 1971.

26. Ibid.

27. Ibid.

28. Edward A. Harris, "The Men Behind McCarthy," *New Republic*, April 24, 1950, p. 10.

29. *Congressional Record*, 82nd Cong., 1st sess., 97, pt. 6: 7703-13 (July 6, 1951); *New York Times*, April 11, 1952, pp. 1, 3.

30. Gouverneur Paulding, "Three Postcards from the Past," *The Reporter*, August 30, 1949, p. 12.

31. Paulding, "The Stars and the Questions," *The Reporter*, August 30, 1949, p. 38.

32. *The Reporter*, September 9, 1965, p. 12.

33. Marya Mannes, *More in Anger* (New York and Philadelphia: Lippincott, 1958), p. 13.

34. Mannes, *Out of My Time*, p. 207.

35. William Harlan Hale, "1948: First Principles," *New Republic,* December 29, 1947, p. 18.

36. *The Reporter*, October 27, 1953.

37. Ascoli contributed a long introduction to the posthumous selection of Berle's papers, *Navigating the Rapids, 1918-1971*, ed. Beatrice Bishop Berle and Travis Beal Jacobs (New York: Harcourt Brace Jovanovich, 1973). Kissinger and Ascoli became friends as a result of discussions following a critical review of Kissinger's book *Nuclear Weapons and Foreign Policy* in *The Reporter* by Paul Nitze. See Bruce Mazlish, *Kissinger: The European Mind in American Policy* (New York: Basic Books, 1976). pp. 109-11.

38. *The Reporter*, October 20, 1955, p. 7.

39. Hentoff's report on black musicians in the United States, "The Case of the Missing Musicians," won him a certificate of recognition from the National Conference of Christians and Jews in 1960.

40. Nat Hentoff, "Speaking Truth to Power," *Village Voice*, June 3, 1965, p. 16.

41. A representative listing of names would include Hanson Baldwin, Marquis Childs, Paul Duke, Julius Duscha, Mel Elfin, John Finney, Max Frankel, Otto Friedrich, Alfred Friendly, Fred Hechinger, Chet Huntley, Marvin Kalb, Stanley Karnow, Spencer Klaw, Eric Larrabee, Anthony Lewis, Herbert L. Matthews, John Mecklin, Walter Millis, R. Hart Phillips, Norman Podhoretz, A. H. Raskin, Chalmers Roberts, M. J. Rossant, Richard Rovere, Daniel Schorr, Sander Vanocur, and Jules Witcover.

42. William L. Rivers, *The Opinionmakers* (Boston: Beacon Press, 1965; reprinted 1967), pp. 89-90.

43. *The Reporter*, April 30, 1959, p. 10. This was the circulation figure Ascoli had hoped to attain when he started the magazine, although the financial break-even point was seen as a circulation of 200,000 with twenty pages of advertising. See *The Reporter*, June 13, 1968, p. 4, and *Current Biography*, February 1954, p. 9.

44. *The Reporter*, April 30, 1959, p. 10.

45. "A Report to Our Readers," *The Reporter*, January 5, 1961, p. 14.

46. ABC statement, June 1966.

47. "A Report to Our Readers," *The Reporter*, January 5, 1961, p. 14.

48. John Fischer, "Editor's Easy Chair: Self-Portrait of the Harper Reader," *Harper's*, September 1958, p. 14.

49. James N. Rosenau, *National Leadership and Foreign Policy* (Princeton: Princeton University Press, 1963), p. 195. The ten magazines cited by participants were *Time* (37 percent), *U.S. News* (24 percent), *Newsweek* (20 percent), *The Reporter* (16 percent), *Foreign Affairs* (15 percent), *Harper's* (9 percent), *Life* (8 percent), *New Republic* (6 percent), *Atlantic* (6 percent), and the *London Economist* (5 percent).

50. These figures are from a study made for *The Reporter* in October 1955 by the market research firm Bennett-Chaikin, Inc.

51. Rivers, p. 31.

52. Interview with Shirley Katzander, January 1971.

53. Robert Lekachman, "Political Magazines," *Book Week*, November 17, 1963, p. 28.

54. Both quotations are from "A Report to Our Readers," *The Reporter*, January 5, 1961, p. 14.

55. Rivers, pp. 54-56.

56. "Publicity Report," 12 pages mimeographed (covering the period from October 19, 1962, to March 31, 1963), from Shirley Katzander's files.

57. *The Reporter*, February 2, 1954, p. 8.

58. *The Reporter*, June 22, 1961, p. 6.

59. Interview with Robert Bingham, January 1971.

60. See Robert Ardrey, *African Genesis* (1961; reprinted New York: Dell, 1967), pp. 186, 188, 201-02.

61. Joseph Wechsberg, *The First Time Around* (Boston and Toronto: Little, Brown, 1970), p. 282.

62. Interview with Irving Kristol, January 1971. Moynihan, who once complained, "I spent a decade, 1955-65, trying to obtain some press coverage of the problem of motor vehicle design, utterly without avail," had apparently forgotten this conspicuous exception to the rejection he encountered elsewhere. See Daniel P. Moynihan, "The Presidency and the Press," *Commentary*, March 1971, p. 50.

63. Advertisements placed by the Foster Parents' Plan in *The Reporter* pulled second only to those it placed in the *New Yorker*. (Letter from Lenore Sorin, associate executive director, Foster Parents' Plan, to Homer L. May of *The Reporter*, October 15, 1964.)

64. Louis M. Lyons, "Reporting in Depth," *New York Times Book Review*, March 6, 1960, p. 10.

65. Ascoli, "Our First Decade," *The Reporter*, April 30, 1959, p. 12.

Chapter Four

This Liberal Magazine

. . . this journalistic disgrace called Reporter.

Senator Styles Bridges, 1954

It is precisely because it is always prepared to call even its own natural friends to task that it stands almost on a plane of its own in our affections.

Senator J. William Fulbright, 1959

What I like best about The Reporter*—even when it infuriates me—is that you call 'em as you see 'em, even when you think an old friend like Senator Fulbright is off base.*

A reader in New York, 1964

The prospectus for *The Reporter* notes at one point that "there are many people in America today who are tired of irresponsible radicalism and of stand-pat, penny-pinching conservatism"; but the logical alternative, liberalism, is not mentioned. Nowhere in the five pages of the prospectus can the words "liberal" or "liberalism" be found. One suspects that there were two reasons for this startling omission. First, the words "liberal" and "liberalism" had been so widely and loosely used that they had become devoid of much real meaning. The prospectus declares war on empty slogans and stereotypes:

> Communism, the most dogmatic of all opponents, can be fought only by free and uncluttered minds. We assume that a large number of Americans are as tired as we are of clichés and stereotypes, and as anxious as we are to have the major problems of our day analyzed and reported in such a way that nothing is taken for granted, not even the righteousness of our democratic position and the wickedness of our Communist opponents. We have set ourselves to be relentless cliché hunters.

The prospectus states that *The Reporter* would be guided by four basic beliefs:

> a) In freedom, which means the capacity that men have of exerting some control over the conditions of their own lives, a capacity that cannot be denied or crippled without making men into the tools of their own destruction.
> b) In America, as a nation whose freedom and well-being are inseparably tied to the freedom and well-being of other nations.
> c) In the function and responsibilities of journalism—provided it meets exact and exacting standards.
> d) And finally, in the I.Q. of the American reader, whose capacity to grasp facts and ideas is crudely underrated by most of the existing media of information.

In restating the first two of these beliefs three years later, Ascoli reaffirmed his dislike of slogans:

> What moves us is a militant faith in freedom and in America. We consider them interchangeable, and we think that this faith can be made much more contagious when attested by one's work rather than by flag waving. Actually we care too deeply for freedom and for America to be flag wavers of "liberalism" or "Americanism."[1]

There was a second reason for avoiding the word "liberal." If it could be virtually meaningless, it could also, in one respect, be misleading. When *The Reporter* began in 1949 there were two long-established and well-known liberal magazines, the *Nation* and the *New Republic.* The former, founded in 1865, called itself the country's "oldest liberal weekly"; the latter, often bracketed with it ideologically–an eager young author once submitted nearly identical articles to each magazine simultaneously in hopes that they would both be printed–was thirty-five years old when *The Reporter* started.[2] Yet a number of liberals were unhappy or uneasy about the variety of liberalism–what Dwight Macdonald had called "totalitarian liberalism"–represented by these two magazines. Ascoli was not alone in his dismay at Michael Straight's selection of Henry Wallace as editor of the *New Republic* late in 1946. Bruce Bliven, long-term editor of the magazine who became "editorial director" during the Wallace regime, soon became convinced that "the American Communist Party had closed in on" Wallace, "doing a superb 'snow job' to convince him that they were just a bunch of good liberals." After Wallace's selection as the Progressive Party's presidential candidate in the summer of 1948, Wallace was asked to resign from the magazine, and the *New Republic* gave its support to Truman. But "the great Wallace debacle," as Bliven called it, which coincided with the magazine's disastrous efforts to convert itself into "a liberal *Time*," did not enhance the reputation of the *New Republic*, and after 1948 its circulation plummeted.[3]

Before long the *Nation* was experiencing problems of its own. Early in 1951 one of its contributors, Clement Greenberg, who had been its art critic from 1943 to 1949, wrote a lengthy letter to the *Nation* complaining that its foreign editor, J. Alvarez Del Vayo, followed "a line which invariably parallels that of Soviet propaganda."[4] The *Nation* refused to print the letter, and Greenberg sent it to the *New Leader.* When that magazine printed it, the *Nation* brought a libel suit–it was dropped after several months–against both Greenberg and the *New Leader.* Robert Bendiner and Reinhold Niebuhr both resigned from the *Nation* in protest against the libel suit, which Niebuhr said "brought to a head my disagreement with the *Nation* on foreign policy."[5] The *New Leader* started a new column, "Letters the *Nation* Wouldn't Print," with contributions from Arthur Schlesinger, Jr., Richard Rovere, and other liberals, protesting the *Nation's* practice of censorship and its attempts to intimidate its critics. The *Partisan Review* supported the *New Leader* and criticized the *Nation* for an obituary of André Gide that the magazine had published in March 1951: "At bottom only pro-Soviet bias can account for a magazine like

The Nation publishing Mr. Werth's calumny of a dead writer whose entire life and work was a search for honesty."[6] That same spring Granville Hicks was analyzing the "pro-Soviet bias" of the eighty-fifth anniversary issue of the *Nation,*[7] and a year later Richard Rovere could still ask rhetorically in the *New Leader* (July 14, 1952), "How Free Is the *Nation*?"

Thus Ascoli, who was to write in 1950, "*The Reporter* is every inch anti-Communist," had reason to try to keep his new magazine from being placed automatically in the same category as the two leading liberal magazines.[8] He wanted it to be judged on its merits. And he wanted, if possible, to rescue the term "liberalism" from the distortions he felt it had acquired:

> For all too long liberalism has represented a depressed and depressing area in the American political landscape, particularly since most of the liberal reforms that had been advocated for a long time have been translated into laws. But a new political struggle is on—not on a national scale but on an international scale—and in this struggle the liberal values need to be thoroughly cleansed of triteness and smugness. They need to reacquire all their assertiveness and buoyancy, for they are the only values that can give safety and peace to the world. *The Reporter* is all out for this militant and buoyant liberalism.[9]

The real shortcoming of the liberal magazines, some of *The Reporter*'s staff felt, was their "doctrinaire" quality. Theodore H. White, who had joined the *New Republic*'s staff during the Wallace period, later said of his time there: "Six months on a liberal magazine will cure almost any illusion. The dogmas were more constricting and binding than those at *Time*."[10] Philip Horton said in 1971, "A refusal to accept the doctrinaire liberal simplifications was really one of the most useful and valuable aspects of *The Reporter*."[11] Ascoli's comments are somewhat more pungent:

> I never objected to the doctrinaire attitude of the *Nation* and the *New Republic*. What I remember having said over and over again was that they were predictable: just by looking at the table of contents of an issue, one could know exactly the position these two magazines were going to take on every single subject they covered. And, pray, where was the doctrine of the doctrinaires? At best, a smudged, fifth carbon copy of the Communist line.[12]

Some former staff members believe that *The Reporter*'s use of slick paper and art work—an expensive way to publish a magazine that initially carried no advertising—was part of a deliberate effort to avoid identification with the two leading liberal magazines, which were printed on cheap stock. *The Reporter*'s publicity releases about its staff in the mid-fifties did not mention that William Harlan Hale had been on the staff of the *New Republic* or that Robert Bendiner had been on the staff of the *Nation*. The omission in the latter case was particularly conspicuous, since Bendiner had been listed as the *Nation*'s managing editor from 1937 to 1944—part of this time he was in the Army—and as its associate editor from 1944 to 1951. Although Ascoli had known Bendiner since the 1930s and had consulted with him about *The Reporter* before it started, he hesitated to invite him to join the staff. Bendiner surmises—correctly, Ascoli says—that Ascoli wanted to avoid too close an identification with one of the existing liberal magazines. "And I think he was right to do this," Bendiner adds; "it was a new magazine, and it needed to be free."[13]

Ironically, however, because *The Reporter* took a vigorous stand that was similar to that of the *Nation* and *New Republic* on two controversial issues—McCarthyism and the China Lobby—it was inevitably associated with them, and like them it was accused of being procommunist. In October 1950, Senator William Knowland denounced *The Reporter* as a "Red" magazine.[14] His wrath had been kindled by the fact that *The Reporter* had sent a packet of six or eight issues to the State Department in the summer of 1950 for circulation among its posts abroad. The packet included the issue of January 3, 1950, which had been devoted to Communist China and had editorially suggested—with important qualifications—that it be recognized by the United States.[15]

The Reporter's famous exposé of the China Lobby in the spring of 1952 did not improve its reputation among conservatives. Senator Knowland insisted that an official of the Communist Party of New York had circulated, early in 1949, a document that had called for "a congressional investigation of the Chinese lobby in Washington."[16] Senator Harry Cain, who submitted what he called "the most documented speech on the question called The China Lobby that has yet been written or spoken on the floor of the United States Senate"—it ran to more than seventy pages—concluded that Ascoli and his associates either had "been 'suckered' by the Communists from start to finish, or they have deliberately participated in a Communist maneuver."[17] In the summer of 1954, after Senator

Styles Bridges' connections with the China Lobby had been set forth in Douglass Cater's lead article, "Senator Styles Bridges and His Far-Flung Constituents" (July 20, 1954), the Senator, up for reelection in November, attacked Ascoli and his staff at length as procommunist.[18]

In May 1953, Ascoli received an anonymous letter from Baltimore, whose sender, apparently a student at the Counter Intelligence Corps School at Fort Holabird, stated that an instructor there had told his classes that *The Reporter* was procommunist. Early in 1954 the *New York Post* noted that the Voice of America had canceled its subcriptions to *The Reporter*, the *Nation*, and the *New Republic*.[19] A few months later Scott McLeod, the State Department's security chief, admitted that he was checking the number of subscribers to *The Reporter* in the State Department—*not* their names, he hastened to add—in order to respond to a Congressional inquiry.[20] And in 1962 Senator Strom Thurmond accused *The Reporter*, together with the *New York Times, Washington Post, Nation, Newsweek*, and *Bulletin of Atomic Scientists,* of following a propaganda line originated by the *Daily Worker*—a line implying "that anti-Communists constituted a graver threat to our country than do Communists."[21] That same year *The Reporter* and the *New Republic* were both banned for classroom use in the high schools in El Segundo, California.[22]

Conservatives were not alone in believing there to be an affinity between the *New Republic* and *The Reporter.* One liberal critic, after identifying the former as being "close to the official organ of the amorphous Kennedy left," added, "Like the *New Republic, The Reporter* sometimes reads more like the liberals' news magazine than like an independent, critical journal of opinion."[23] With the advent of the New Frontier, both magazines had a marked increase in circulation. And in 1964 when the *New Republic* conducted a readership survey it found that the other magazine most frequently read by its readers was *The Reporter.*[24]

Thus at least until the early 1960s *The Reporter* was often popularly associated with the long-established liberal magazines. In the mid-1950s, in fact, some observers considered it *the* liberal magazine. A liberal admirer —and later a contributor—Alastair Buchan of the *London Observer*, called it in 1953 "the most trenchant liberal magazine in the United States."[25] A conservative opponent, Willmoore Kendall of the *National Review,* called it in 1956 the "current locus classicus for the Liberal line."[26]

Such statements cannot necessarily be taken at face value, for Ascoli's conception of *The Reporter*'s liberalism often differed from the conceptions of *The Reporter*'s liberalism that were held both by his supporters and his critics. Because *The Reporter* agreed with the other elements of the liberal community on three of the major liberal issues of the 1950s—civil liberties, civil rights, and nuclear testing—its deviations from the then prevailing liberal thought went largely unnoticed. The two chief deviations—a distrust of big government and a persistent anticommunism—became more noticeable in the 1960s, when Joe McCarthy had passed into history, the Limited Test Ban Treaty of 1963 had been signed, and an outspoken anticommunism had come to be considered somewhat passé. ("We're both liberals of the fifties," Irving Kristol said in 1971 of himself and Ascoli.)[27] "Hubert Humphrey, Walt Rostow, Paul Douglas, John Roche, and Max Ascoli have kept the faith," Richard Rovere said in 1967, adding, "I do not think it a faith worth keeping."[28]

Although *The Reporter*'s position on such noneconomic issues as civil liberties, civil rights, and nuclear testing was clearly in the mainstream of postwar American liberalism, on economic issues it was less predictable. The prospectus stated: "*Labor*, in our opinion, is the most momentous of our national problems. . . . *The Reporter* is going to cover the field of labor steadily, as a friend of labor. To be a friend, of course, does not mean to be a flatterer."[29] A number of Paul Jacobs' exposés in the 1950s—"The Labor Movement Cripples a Union" (November 1, 1956), "The World of Jimmy Hoffa" (January 24 and February 7, 1957), and "The Negro Worker Asserts His Rights" (July 23, 1959)—were hardly flattering to important elements in organized labor, but as *The Reporter* commented when it printed the first of these, "It is precisely because we are friends of labor that we do not think discussion of what is wrong with it should be left to the Peglers."[30] Instead of praising or defending "labor" as an abstract entity, Jacobs' articles normally displayed compassion for the plight of working people, as is well illustrated by his study of migrant farm workers, "The Forgotten People" (January 22, 1959). After Jacobs left *The Reporter* in 1960 its labor coverage changed in tone. The *New York Times*'s highly regarded labor specialist, A. H. Raskin, contributed a number of thoughtful, judicious pieces such as "Civil Rights: The Law and the Unions" (September 10, 1964), "Labor's Political Frustrations" (April 7, 1966), and "Labor's Middle-Class Revolt" (June 15, 1967). Raskin's articles had perhaps a better claim to intellectual respectability, but Jacobs' articles had seemed more readable.

Ascoli could never work up much enthusiasm for writing editorials on economic questions—the prospectus had said that the magazine would be "quite pragmatic when it comes to economics"—probably because he did not have a strong ideological position on economic issues. The prospectus had promised that *The Reporter* would "never lose sight of what happens to the individuals in every concrete situation we try to report," and the magazine did not hesitate to cover the problems of the poor and of minorities. But it was reluctant to advocate massive government action as a panacea, for Ascoli had an abiding distrust of big government—a distrust that many liberals, by the end of the 1960s, came to share.[31] As fervently as he defended President Johnson, he admitted that he did not much care for the Great Society. Ascoli stated the dilemma well—it is the continuing conflict between economic and noneconomic liberalism—in an early editorial:

> There is nothing more legitimate than the demands of special groups, or the nation as a whole, for increased social and individual security, but there is a great deal wrong when, to satisfy these demands, a powerful Federal bureaucracy stifles group initiative and local pride. . . .
>
> Freedom is not maintained when the livelihood of the citizens and their protection from the major accidents of life depend entirely on omnipotent officials no matter how good their intentions may be—just as we kill freedom when we cause destruction that we cannot repair. To remain free, we must never give any individual powers that are beyond men's measure, nor must we ever destroy more than we can rebuild.[32]

Ascoli in this respect appeared out of phase with liberal opinion. Just as his anticommunism persisted too long—or so some of his critics would maintain—so did his criticism of big government begin some fifteen years before it could be considered respectably liberal.

CIVIL LIBERTIES

The Reporter's single most valuable contribution was in the area of civil liberties. Several months before the emergence of Senator Joe McCarthy, Ascoli had devoted an issue of the magazine to civil liberties and had shown himself to be, in one respect, a poor prophet. Finding no "Grand Inquisitor" among the security investigators, Ascoli wrote: "It will be difficult, when history is written, to label this phase of American life with the name of any one man—a singular evidence of how scarce both heroes and villains have become on our national scene."[33] When a budding Grand Inquisitor did make his appearance in February 1950, however, with a speech in Wheeling, West Virginia, Ascoli was quick to recognize him for what he was, and *The Reporter* was the first national magazine to devote a special issue (June 6, 1950) to the menace of McCarthyism.[34]

The Reporter was not alone in making an early attack on McCarthy. The *Nation* and *New Republic* had each lashed out at him editorially in March 1950. In April the *Saturday Review* had commented editorially on "the Senator's reckless conduct." The *New Yorker*'s first notice of the Senator came with Richard Rovere's "Letter from Washington" in May. *Time*—in marked contrast to *Newsweek*, which treated McCarthy like any other weekly news sensation—was early in its news stories to call him "an irresponsible demagogue."[35] *The Reporter*'s first mention of McCarthy came in "The Reporter's Notes" in the April 11 issue. It was a brief satirical item that imagined a Soviet bureaucrat's gratitude to McCarthy for proving that "some of the best things that happen to Russia are not the result of *Politburo* planning." The next issue contained an article written by a reporter on the *Milwaukee Journal* that described a number of the Senator's questionable activities in Wisconsin politics. Calling him "one of the most cynical men I have ever known," the author wrote, "McCarthy's current behavior in Washington is in line with past performances."[36]

Then Ascoli decided that the story was important enough to merit a special issue of the magazine:

> I remember we were preparing an issue on education, and ten days before closing—it threw everybody for a loop—I said, "Throw out education. Let's write about this man and the victims he has made. I want to know more about Lattimore. I want to know more about all these people. The danger has to be denounced."

As one former staff member recalls this editorial meeting:

> Ascoli said, "This is a terrible and important thing that we have to do something about." And there were people then on the staff

> who said, "Well, yes, it's bad, but you know. . . . It's only some jerk from Wisconsin. It's not a big thing. Let's not get too excited about it." And I clearly remember Ascoli being very insistent. He said he'd seen things like this in Italy, and the pattern was all too familiar to him.

Ascoli later recalled his opposition to McCarthy as having been too early to be entirely respectable. Fifteen years later, in an editorial defending President Johnson, he commented on one unfortunate result of McCarthy's antics, together with a scornful glance at those who had been slow to oppose him:

> When the President acted in Santo Domingo the uproar was all the greater because McCarthy had discredited anti-Communism. It was because of McCarthy that the fear of Communism first spread a blinding, thoughtless obsession that actually helped Communism assume all manner of safe disguises. Even now it is impossible to mention any given number of Communists without evoking the memory of the man who had in his hand the irrefutable evidence of ever-changing numbers. At present some of the communicators who were not too prominent in the fight against McCarthy feel that they have been given a second chance.[37]

The June 6, 1950, issue of *The Reporter* opened with a seventeen-page section headed "Behind and Beyond McCarthy," which was an analysis of the conditions that allowed such a demagogue to gain power and included an article by William H. Hessler, "Ordeal by Headline"; Claire Neikind's "U.S. Communists—1950," which argued that the numbers and influence of the U. S. Communist Party were on the decline; Stuart Long's "Thunder on the Right," an account of ultraconservatives in Texas; a staff-written article, "On Reading Lattimore"; Jean-Jacques Servan-Schreiber's "How It Looks from Europe"; and Douglass Cater's "The Captive Press," which showed how McCarthy used the weaknesses of the press to capture the headlines. McCarthyism gives rise, wrote Cater, "to fear of what a demagogue can do to America while the press helplessly gives its unwilling cooperation."[38]

The article "On Reading Lattimore" illustrates well *The Reporter*'s penchant for making independent judgments. Noting that neither McCarthy nor his supporters had claimed to have read the writings of Owen Lattimore, McCarthy's chief target, the article sums up the reaction of the magazine's research staff after they had read all of Lattimore's available works. The final two sentences of the article convey its tone: "From the printed page, Lattimore appears intense, everbusy, and sometimes pompous—a man who sometimes had seemed to swallow Communist slogans, but who has industriously tried to present the truth as he saw it. This ebullient publicist is as unfit for the dark role of Soviet architect of U. S. foreign policy as Senator McCarthy is for the U. S. Supreme Court." Lattimore expressed gratitude for *The Reporter*'s defense of him, but, understandably, was not entirely happy with the way in which he had been defended.[39]

Ascoli's editorial in this issue, "The G.O.P.'s Choice," is worth noting for two reasons. First, having learned in Italy that mere diatribes against a demagogue—however soul-satisfying—do not dislodge him from power, Ascoli does not engage in polemics against McCarthy, whom he seems to regard as almost beneath his contempt. Instead, he calls on the Republican Party to repudiate "the youthful anarchist from Wisconsin." (Late in 1952 Ascoli described himself as "a frustrated Republican; I would like nothing better than to go along with the liberal wing of the G.O.P., but I am seldom if ever given a chance.")[40]

Second, the editorial illustrates how Ascoli is unable to discuss any political question in purely domestic terms:

> Now Senator McCarthy seems to have tapped, quite accidentally, a reservoir of long-hidden malice. He has offered to the American people a chance to escape the rigors of the world civil war by fighting their own civil war against the McCarthy-made Communists at home. . . . This is happening exactly at the time when we most urgently need a strong opposition party, capable, if it is returned to power, of leading the nation in the global conflict against Communism. In our days, the constituency of the American government has become much larger than the American electorate. The nations on our side are so dependent on us that without our assistance and advice they could not survive. As far as we are concerned, if these nations should move or be dragged into the Communist camp, our own survival would be endangered. . . .
>
> The function of our major political parties today is to link the local, sectional, and occupational interests of the American

people to the broader international structure, of which America
is the decisive, but by no means the only, part.

McCarthyism continued to be one of *The Reporter*'s major targets. Senator Millard Tydings contributed "McCarthyism: How It All Began" to the issue of August 19, 1952. The issue of July 21, 1953, under the heading "The Purgers and the Purged," ran an editorial by Ascoli urging the President to repudiate McCarthy; Richard Rovere's "The Adventures of Cohn and Schine"; a personal account by Theodore Kaghan, who had been forced by McCarthy to resign from his position as acting deputy director of the Public Affairs Division of the U.S. High Commission for Germany, "The McCarthyization of Theodore Kaghan"; Philip Horton's account of McCarthyism within the Voice of America, "Voices within the Voice"; and Raymond Swing's "V.O.A.—A Survey of the Wreckage." The following spring *The Reporter* published Anthony Lewis' account of the ordeal of a Navy civil servant, Abraham Chasanow, "What Happens to a Victim of Nameless Accusers" (March 2, 1954), a classic account of the evils of the McCarthy era. It helped Chasanow—called "Bernard Goldsmith" in Lewis' account—win back his job; it won for Lewis the Heywood Broun Award and for *The Reporter* a Benjamin Franklin Magazine Citation for "meritorious contribution in the field of public service."

But while emphasizing the costs of McCarthyism in individual lives, *The Reporter* also tried to see the phenomenon in historical perspective. The issue of September 14, 1954, contained an article by Eric Goldman, "There's Nothing New About McCarthyism," and the first installment of a two-part article by Theodore H. White, "U. S. Science: The Troubled Quest," an examination of the reactions of American scientists to the government security system in the post-Oppenheimer era. The White articles elicited from Albert Einstein the most famous letter that *The Reporter* ever received—it rated a front-page article and an editorial in the *New York Times*—a short letter in which Einstein said that if he were a young man he would become a plumber rather than a scientist or scholar, in order to have some independence. In an editorial note following the letter, Ascoli wrote that both the far left and the far right would quote the letter gleefully, but added, "The publication of this letter—and indeed of our magazine—is an act of faith in the sanity of America."[41] Three years later *The Reporter* published an article by Giorio DeSantillana, "Galileo and J. Robert Oppenheimer" (December 26, 1957). Saul Padover, in a letter to the editor, wrote that the article exemplified "*The Reporter*'s basic approach as I see it; namely that for an understanding of politics, history properly used casts the steadiest light, and that in fact without a historical perspective contemporary events seem merely sensational or meaningless."[42]

Part of *The Reporter*'s battle against McCarthyism was the opening of its pages to some of McCarthy's victims. Edward Posniak, the Moscow-born economist who had been accused by McCarthy of being a Communist "Mr. X," contributed several articles, as did Charles W. Thayer, a career Foreign Service officer who had been dismissed by Scott McLeod. Martin Merson, a businessman who had served several months as a consultant to the head of the International Information Agency—a predecessor of the United States Information Agency—wrote a long account of his battles there against McCarthy, "My Education in Government" (October 7, 1954). Edward Corsi, a former commissioner of immigration who had been invited by John Foster Dulles to expedite the admission of immigrants under the Refugee Relief Act of 1953 and had then been summarily dismissed after attacks from conservatives, told his story in "My Ninety Days in Washington" (May 5, 1955).

The Reporter served not only as a forum for dissenters against McCarthyism at home but also as a forum for dissenters against repressive regimes abroad. The issue of January 22, 1952, contained an article adapted from a speech given in Paris by Czeslaw Milosz, the Polish poet who had fled to the West and who later wrote the sensitive study of East European Communism, *The Captive Mind.* The same issue that contained William Harlan Hale's uncomplimentary article on Scott McLeod (August 17, 1954) also contained an article, "Chinese Co-ed in a Communist College," by a refugee who wrote under the pseudonym of "Marya Yen" because her family in China was exposed to reprisal. Shortly after the Cuban missile crisis *The Reporter* published two reports from inside Cuba by a pseudonymous Cuban correspondent, "Panatela" (December 6, 1962), and March 13, 1963).

In the summer of 1958 *The Reporter* published three long excerpts from Boris Pasternak's *Doctor Zhivago*—it was the first American magazine to do so—shortly before the book's publication in English. Ascoli did not regard the book as anticommunist; that, he believed, was too simplistic and misleading a label. Rather, he saw the book as exemplifying one of his most persistent convictions—one that underlies all of his political writings and was an unstated premise in his debate in 1956 with Schlesinger: "The quietly stated message of the book is that politics cannot control the whole of life, for there is in man an inner realm that no politics can touch."[43] When he printed several paragraphs from *Doctor Zhivago* as the Christmas editorial for 1958, Ascoli again stated his view of Pasternak's theme: "His novel is, and will remain, eternal dissent from the legitimacy of totalitarian politics—from all efforts to make man into a merely political animal."[44] Later *The Reporter* published stories by other dissident Soviet writers: Vladimir Dudintsev (May 26, 1960); "Nikolai Arzak"—pseudonym of Yuli Daniel, whose trial was reported by George Bailey in 1966 (August 16, 1962); Fyodor Abramov (January 20, 1963); and Lazar Lagin (November 19, 1964).

Although its vigorous support of free speech at home and abroad was *The Reporter*'s most conspicuous contribution in the area of civil liberties, its concern about invasion of privacy was reflected in two exposés by the investigative team of William S. Fairfield and Charles Clift, "The Wiretappers" (December 23, 1952), and "The Private Eyes" (February 10, 1955). Its concern for separation of church and state was reflected in an item in "The Reporter's Notes" (July 19, 1962) welcoming the Supreme Court's decision on school prayers and in an article by Washington lawyer Marcus Cohn, "Religion and the FCC" (January 14, 1965), arguing that the FCC's policy on religious programming was contrary to recent Supreme Court decisions.

CIVIL RIGHTS

Civil rights was not among the handful of issues named by Ascoli in his tenth-anniversary editorial or in his final editorial, in both of which he listed the issues on which he was proud to have taken a stand. And *The Reporter*'s stand on civil rights was not a conspicuous one, as it was on McCarthyism and the China Lobby. Nevertheless, from the early months of the magazine's existence, the plight of the blacks in America received extensive coverage, and Ascoli repeatedly called for full recognition of their civil rights. An issue devoted to the theme of "The Negro Citizen" (December 6, 1949) came respectably early,[45] and Bishop Bernard J. Shiel of Chicago wrote a few weeks later, "Your commonsense approach, in recent issues, has been of no little help to us in our thinking about FEPC, for example, in the State of Illinois."[46]

The Reporter's "commonsense approach" to civil rights could be said to reflect the magazine's Credo ("Belief in the Freedom of the Individual), Tenet ("Responsibility: It is not enough to criticize; one must always offer an alternative"), and Maxim ("Always Objective; Never Impartial"), which were proclaimed on a wall plaque in Ascoli's office. The approach had three characteristics: an emphasis on practical results rather than on name-calling; a belief in the right to vote as the key issue that would enable blacks to fight effectively for other rights; and an insistence on seeing the racial issue in its largest dimensions—that is, as a national problem rather than merely a sectional one, as a human problem rather than merely a legal one, and as a global problem rather than merely a domestic one.

Although *The Reporter* carried a number of articles in the 1950s dealing with racial problems in the North or on a national scale,[47] its chief concentration naturally was on what it called in the issue of March 8, 1956, "The Ordeal of the South." In that issue Ascoli's editorial, which was an endorsement of Adlai Stevenson's stand on civil rights, stated his own position in its opening paragraph:

> What a national tragedy this whole thing is, and how humble we feel when we think of those anxious fellow citizens down there in the South, white and colored, who are still trying to keep communications open between the two races. But that humility, intense as it is, brings with it an intolerance just as intense of those synthetic John Browns who, comfortably located in the North, contemplate no Harper's Ferry and are so willing to shed their ink. We allude to the brave politicians and gallant editorialists who are howling for Federal intervention. As for the pale imitators of Calhoun and the revivers of the Klan, we can only say that they are beneath contempt.[48]

Readers' reactions were mixed. Two of them—one from Kansas and one from New York—praised him for his "eloquent plea for moderation." A black reader in Philadelphia was disappointed in "what I believed to be a staunch, liberal, and militant magazine." He found the editorial and accompanying articles to "smack of but one thing: appeasement." Another reader commented, "While I think that *The Reporter*'s counsel of prudence and moderation in the crisis of integration is highly commendable, it seems to me that such an attitude can too easily put us to sleep."[49]

The Reporter continually tried to get beyond abstractions and see the human dimensions of the racial problem. During the 1950s Douglass Cater, a Southern liberal who had gone north from Alabama, and David Halberstam, a Northern liberal who had gone south to Mississippi and Tennessee, both reported for the magazine on racial conditions in the South. William Demby, a black writer traveling through the South, wrote " 'They Surely Can't Stop Us Now' " (April 5, 1961). C. Eric Lincoln contributed "The Strategy of a Sit-in" (January 5, 1961). An article by Bettye Rice Hughes, "A Negro Tourist in Dixie" (April 26, 1962), prompted Gordon R. Carey, program director of the Congress of Racial Equality, to write: "*The Reporter* has performed a valuable service by publishing this article. It should encourage the timid to assert their rights and their dignity."[50] Vincent Harding, a black lay minister from New York, contributed "A Beginning in Birmingham" (June 6, 1963) and the issue of March 25, 1965, carried a dialogue between Ralph Ellison and Robert Penn Warren, a prepublication excerpt from Warren's thoughtful book *Who Speaks for the Negro?*

The problems of reporting the emotion-laden issue of race are illustrated by the publicity that resulted from the famous interview of William Faulkner by Russell Warren Howe, which appeared in the March 22, 1956, issue of *The Reporter.* Faulkner's remarks are those of a sensitive, proud, and intelligent Southern white who recognizes the psychological problems his people face in giving long-overdue justice to blacks; the interview drew thoughtful comments from *The Reporter*'s readers, many of whom were impatient with Faulkner's "moderate" or "mystical" opinions. But when *Time, Newsweek,* and the *New York Times* gave their reports of *The Reporter* interview, they quoted little except one or two of Faulkner's melodramatic remarks on Southern resistance, and Faulkner protested against the distortion of his views.[51] The difference between the interview itself—disturbing but thought-provoking—and the reporting of it at second-hand—sensational and emotion-producing—illustrates the distinction that Ascoli felt existed between his magazine and the magazines of "news and opinion."

Although Ascoli recognized that the blacks "have certainly greater and meaner handicaps to overcome than all the other racial groups that form America,"[52] he believed that when blacks in the South enjoyed the franchise they could then fight politically for their rights as other ethnic groups in America—Irish, Germans, Italians, and Jews—had fought.[53]

He argued that Southern politicians would "change their attitude toward the Negro problem only when they are compelled to take into account the Southern Negro vote."[54] Robert Bendiner's analysis, "The Negro Vote and the Democrats," appeared in the issue of May 31, 1956. A decade later an article by Reese Cleghorn and Pat Watters, "The Impact of Negro Votes on Southern Politics" (January 26, 1967) concluded optimistically:

> The hopeful aspects of the Southern political situation . . . are to be found in the fact that in 1966, Negroes and whites were moving toward a working political opposition to the forces of racism in every Southern state. This was not true ten years ago, or even two.

Ascoli recognized the urgency for civil-rights legislation, but unlike some of his fellow liberals he believed that legislation needed to be accompanied by restraint and charity—or what he thought should properly be called "love":

> Love does not mean refusal to pass judgment on those who are guilty of evildoing or to grant indiscriminate amnesty on the ground of ancestral guilt. Love means active faith in the ineradicable humaneness even of those who never knew they had it buried within themselves. Strangely enough, the clergy in our time, and most distinctly about the Negro problem, seem to have been more concerned with revindications and causes than with the quiet, unceasing, soul-stirring search for love.
>
> As far as our Negro brothers are concerned, part but only part of the long work on their behalf can be done by public institutions. The other part, perhaps the essential one, can only be accomplished by man-to-man co-operation. There is the danger of legislative inflation—too many hastily passed laws minutely dictating the behavior of Negroes and whites alike. It is only when legal or moral principles are filtered through the conscience of the individual that they acquire a truly seminal quality.[55]

A few months earlier *The Reporter* had run two contrasting case histories from Harlem—George R. Metcalf's "Metro North Moves Mountains" and Rasa Gustaitis' "The Angry Parents of I.S. 201"—to illustrate "what can be accomplished when the people of a decaying neighborhood are helped to help themselves, and what can be lost when a bureaucracy grinds ahead without enough regard for the human element."[56]

Finally, Ascoli insisted on seeing America's racial problems in a global context. After the school crisis in Little Rock, he pointed out sadly that the influence of his friend Senator Fulbright as "conscience" of the Senate and as watchman of United States prestige in the world had been inevitably debased by Fulbright's plea for delay in the integration of Central High.[57] In a later editorial he deplored the tendency for America's racial crisis to encourage a masochistic and morbid isolationism:

> They keep recurring, these gloomy warnings, whenever things occur within our borders that make men of good will ashamed: let's put our house in order, it is said, before we concern ourselves with the disorder that prevails in distant lands. . . .
>
> A fleeting mood of despondency may lead to such yearnings for a Little America, virtuous, self-isolated, and frugal. But we cannot let ourselves long be taken by such fancies without losing the sense of our institutions, and of our country. . . .
>
> We are a great democracy with a destiny of its own, equally and inextricably concerned with the sections of the world, and of our own people, that are still undeveloped.[58]

Ultimately, then, America's domestic racial problem was another question that Ascoli could not view in isolation from its effect on America's relations with the rest of the world.

NUCLEAR TESTING

A third liberal issue, nuclear testing, has obvious international implications. "There Must Be an End to It" was the heading of a strong editorial Ascoli wrote for the issue of May 16, 1957, which contained Paul Jacobs' powerful exposé of the fallout from nuclear tests in Nevada. The April 3, 1958, issue contained excerpts from Harrison Brown's lectures urging a limit on atomic tests; a year later Senator Frank Church contributed " 'We Must Stop Poisoning the Air.' " (April 16, 1959); and the July 9, 1959, issue contained an article by Walter Schneir, "Primer on

Fallout." Ascoli was critical of a series of U.S. atmospheric tests in 1962 and disturbed by President Kennedy's apparent flippancy about their effects, but admitted that "the latest Soviet tests left him no choice" and concluded that "we think we understand our President and we stick by him."[59] Yet his reactions to the Limited Test Ban Treaty of 1963 were mixed. He welcomed it because it did promise to end the atomic tests that contaminated the atmosphere, but he nevertheless described it as a "bad treaty" because he feared it could be widely misunderstood as signaling a détente that did not really exist.[60] On the whole question of military preparedness Ascoli did not escape the seeming inconsistency shown by many liberals in the 1950s: while blaming President Eisenhower for not doing more to reach an agreement on nuclear testing—an argument with pacifistic overtones—they simultaneously attacked him for allowing an alleged "missile gap" to develop—an argument with bellicose overtones. Ascoli's editorial "Thank You, Sputnik!" (October 31, 1957), which attacked the administration for its "missile gap," followed by less than six months his strong protest against nuclear testing.

DEFINING LIBERALISM

The Reporter's position on civil liberties, civil rights, and nuclear testing, then, justified its readers in regarding it as a leading exponent of liberalism. Yet Ascoli was aware from the beginning that *The Reporter*'s liberalism was a distinctive variety, and he continually attempted to define it. One early statement leans toward whimsicality:

> The question comes up over and over again: *The Reporter* is a liberal magazine, all right, but what is its brand of liberalism?
> We have an answer: Liberalism without tears.[61]

A few months later, Ascoli gave a one-sentence description of *The Reporter*: "It is a magazine for the partisans of freedom."[62] The most explicit statement of *The Reporter*'s liberalism is Ascoli's editorial "This Liberal Magazine" in the sixth anniversary issue (April 21, 1955) which was reprinted in both anthologies, *The Reporter Reader* (1956) and *Our Times* (1960). After noting the magazine's belief in liberalism, Ascoli points out that few words "have been so blurred by the 'yes, but' treatment."

He notes that the idea of liberty has become "a disembodied principle so secure in the high heaven of abstraction as to require no effort to be made operational." "Yet what is a liberal," Ascoli asks, "if not a man who gives all he has to make liberty operational, and develops the highest possible degree of skill this vocation demands?"

"*The Reporter*'s liberalism," he adds, "is based on the belief that liberty, far from being an ethereal thing, is always identified with and related to specific and present situations." Thus:

> A liberal, a man who cultivates the skills that make freedom operational, is always a man on special assignment.
>
> Because of its devotion to freedom, this magazine is always on special assignment. . . .
>
> News and opinion is the journalistic currency—frequently manipulated—by which facts and ideas are given circulation. Liberalism means an unending search for the right operational balance and, in the case of this publication, this means the right balance between facts and ideas.

It is easy to see why the political philosopher felt compelled to plunge into journalism. But the political philosopher keeps reasserting himself, and Ascoli complains that American liberalism

> has been singularly leery of defining its own theoretical principles, the set of ideas by which its operations are guided. This has greatly contributed to the lack of commonly accepted standards even in the most articulate liberals.
>
> Liberalism in our country has developed into an instinct, to be sure, exemplified by some of the noblest characters in political life. But this particular condition has been a very heavy handicap on liberalism lately, when it has had to withstand two different yet equally demanding tests. The first was the assault of ruthless, seditious demagoguery. The second was the task of presenting the American case to the outside world. Useful as they are, our political parties cannot do this job. It can be done only by liberals, assigned to this particular task. So far, unfortunately, the language of American liberalism has proved to be strictly for internal consumption.

In the final paragraph of the editorial Ascoli again insists on the necessary limitations of politics: "In one respect, *The Reporter* has not lived up to its liberal creed. It still has too much politics. For a liberal, approximately only half of a man is a political animal–Caesar's half."

Nine years later, on *The Reporter*'s fifteenth anniversary, Ascoli published an editorial headed "This Liberal Magazine, 1964" (May 11, 1964). Here he is more concrete than in the earlier editorial, and the dominant issue is America's relations with the rest of the world, particularly with the Communist nations. The McCarthy era, Ascoli recalls, was "a nearly perfect black-and-white state of affairs." It was a period when the inclination "to look for an intramural, nation-wide escape from the confrontation with international Communism reached its most lurid form." Without obvious villains like McCarthy, Ascoli admits, "It has been far more difficult for a liberal to do his job during these last few years." One of the chief difficulties is the disturbing tendency he finds among some liberals to regard the "basic immorality" of Communism as mythical:

> And if we say that the evil of Communist states is a myth, then so is freedom and so is America.
>
> A liberal cannot give up his concern with ethics, for ethics is about half of his world. He must know how to deal with ethics and with politics and be respectful of both in the concrete situations in which he works. The ultimate test of ethics is not in an empyrean realm but in what happens to the individual in political societies, and most particularly in those where the autonomy of the individual conscience is banished. That autonomy, by definition, cannot be equally enjoyed by all; but when it is prohibited to all, that very prohibition becomes the evidence of evil.

Ascoli finds that the acquisition by America of ultimate power has brought a new temptation: to play God to other nations. "But," he insists, "the way out cannot be found by playing co-god with the Russians. Now and forever, we are stuck with our ethical principles."[63] Four years later Ascoli wrote: "The American liberal . . . has succumbed too frequently of late to a delusion about our external opponent's peaceful intentions" and added that "the present leadership of the Americans for Democratic Action is not representative of American liberalism."[64]

In his final editorial in *The Reporter* Ascoli again insisted that ethics was central to his conception of liberalism:

> But practically everybody is saying now that it is pedantic and old-fashioned to insist on ethical principles when it comes to political regimes, most particularly those of the Left, and to the behavior of the young, especially when uncontaminated by ideals. This bypassing of ethics has been the major heresy of our times.
>
> Has *The Reporter* done well enough in denouncing and fighting back these most amoral trends? I doubt it. In the assignment it has given itself it has been endeavoring to carry through a type of liberalism different from the amorphous one prevailing in this country—a strange mixture of old-fashioned populism sustained by the firm belief in the healing power of bureaucracy, particularly when this power is exercised by freewheeling liberals. No, *The Reporter* hasn't done enough in counteracting the most weedy strains of liberalism. On a number of occasions it did call a few of these liberals by their right name. But it did so too early, thereby offending them twice; once when the description was made, and then when they had to accept it.[65]

Philosophically, few liberals would disagree with Ascoli's insistence on ethics—although many would argue that opposition to the Vietnam War was more ethical than support of it—and tactically, few would disagree with Ascoli's stands on civil liberties, civil rights, and nuclear testing, although some might dissociate themselves from the philosophical reasons he gave for his stands. Politics does indeed make strange bedfellows: those who profess agreement on abstract principles may frequently find themselves opposed on a specific issue; and those who join in support of a particular issue may do so for quite diverse reasons. But politics is also the interaction of personalities, and further dimensions of *The Reporter*'s liberalism can be seen in its relationship with the four successive presidential administrations between 1949 and 1968.

THE REPORTER AND FOUR PRESIDENTS

Harry Truman was President when *The Reporter* started, and Ascoli says, "The president with whom I have had the closest personal relation, strange to say—me, every inch an intellectual—was Truman." He was, in fact, "the only president whom I really liked." Shortly after President Kennedy's inauguration, Ascoli wrote: "Probably our country has become incurably addicted to the father, or in this case, the son image"—a most unhealthy contrast to Truman, "whose earthy, humorous plainness belied any possible attempt at mythmaking."[66]

In November 1951, *The Reporter* became one of the first publications to come out for Eisenhower, and the reason for its support, as described four years later, reflects *The Reporter*'s strong internationalist position: "As head of our nation, we thought, he could at the same time represent the American people and the broader Allied constituency that has no vote but whose survival is interwoven with ours."[67] But in mid-September 1952, when the magazine made its famous volte-face and came out for Stevenson, Ascoli was quoted by the *New York Times* as saying, "Since Eisenhower's return from Europe, we, together with a very large number of other citizens, have not been able to recognize in him the leader we expected."[68] Specifically, he had abandoned internationalism and had surrendered to Senator Taft on foreign policy, as Ascoli and some thirty other liberals charged in a letter to the *New York Times.*[69] When Eisenhower "set himself to harmonize the G.O.P. and took a my-party-right-or-wrong position," Ascoli wrote in 1955, "he was driven to crusade against fellow Americans whose patriotism he had no right to question, and suffered in public promiscuous association with men beneath his contempt."[70] A later private account by Ascoli contains vivid details that did not appear in his public statements at the time:

> It must have been in late August of '52. My wife and I had been looking at a Republican TV program. It showed Ike campaigning in the Middle West, going around arm-in-arm with Senator Jenner—the man who had called George Marshall "a living lie." I had a terrible night, and in the morning asked my wife to look at my back, which was itching all over. Briefly, I had come down with shingles, and out for Stevenson. The shingles were no fun, and took some time to get rid of. In the meantime, I heard that a British friend, Joe Miller, correspondent of the *London Times* who had been on the Eisenhower train, had had an attack of shingles. I wrote him, but unfortunately we concluded that the thing had not acquired epidemic proportions. Anyway, as soon as I had recovered, I went to see Stevenson in Springfield. The first

person I met in the corridor of the Lincoln Hotel was Ken Galbraith. "What are you doing here?" he asked. "Ken," I said, "I have come to join the losing party."[71]

After Eisenhower had become President, Ascoli strongly criticized the "religious verbiage" in some of his speeches, characterizing it as "bureaucratized, routinized godliness" and "vapid, ersatz religion" which "has gone so far as to border on blasphemy."[72] Repeatedly he accused the President of "indecisiveness" and lack of leadership. Eisenhower mirrored, rather than led, the nation:

> In the Chief Executive the nation finds reflected its present mood and some of its most deep-rooted traits: fundamental decency, an earnest wish to do the right thing, a nostalgia for the moral and practical standards that prevailed in the good old days.[73]

Yet at the same time Ascoli spoke warmly of Eisenhower's inaugural address in 1953, of his fight against the Bricker Amendment, and especially of his leadership at the Geneva Conference in 1955:

> There is an element of greatness about him, a power for creating some kind of unity and harmony among people he is associated with, no matter how broad the association may be. Indeed, it may be said that the broader his sphere of action and the greater the multiplicity of elements, the more effectively the peculiar talent of Ike Eisenhower functions.[74]

After Eisenhower's farewell address, Ascoli again deplored his reluctance to use his power and prestige while in office but praised him for his warnings about the "military-industrial complex" and the "scientific-technological elite": "In a peculiar way this man is a liberal, and as a true liberal he is alert to the danger that our freedoms may be curtailed by the organizations established to defend them."[75]

Ascoli rarely found anything to praise in the speeches or actions of Eisenhower's Vice President. A *Reporter* correspondent, Richard Donovan, was one of the three journalists who uncovered the story of the Nixon fund in 1952.[76] In the election year of 1956 William Lee Miller wrote "The Debating Career of Richard M. Nixon" (April 19, 1956), a forerunner of the most famous among the many *Reporter* attacks on Nixon, Meg Greenfield's tour de force "The Prose of Richard M. Nixon" (September 29, 1960), which appeared just before the 1960 election.[77]

During Eisenhower's two terms, then, *The Reporter* was normally found among the opposition. It frequently opened its pages to Democratic congressmen and senators, including John Kennedy, who submitted an article, "When the Executive Fails to Lead" (September 19, 1958), and a book review of General James Gavin's *War and Peace in the Space Age* (October 30, 1958). But despite *The Reporter*'s opposition to Eisenhower and its endorsement of Stevenson in 1952, it was slow to endorse Stevenson for 1956. The issue of December 29, 1955, carried a two-page editorial, "Dear Governor Stevenson," in which Ascoli urged him to prove himself worthy of support. "The major, the all-important test," wrote Ascoli, "will be on foreign affairs. . . . Will you prove . . . that foreign affairs have first, second, and third priority?"

The somewhat patronizing tone of the editorial—and *The Reporter*'s reluctance to jump on the Stevenson bandwagon—brought an editorial reprimand from the *New Republic.*[78] To any outsider, it would seem logical that Ascoli should jump to support Stevenson. The two men had met at the end of 1943, and Stevenson wrote glowing fan letters to *The Reporter* on its big anniversaries. Ascoli had contributed $500 to Stevenson's campaign for governor of Illinois in 1948, and when the full-page list of contributors appeared in the *New York Times* on September 28, 1952, Ascoli's picture was one of the three that appeared on the page. (One reader of *The Reporter* was prompted to inquire whether the magazine's original support of Eisenhower had been genuine; he was assured that it had been.) Privately, however, Ascoli was more critical of Stevenson than he felt he could be in public, as appears in his later personal account of their relationship:

> In the early summer of 1944, Stevenson came to our country house in Croton-on-Hudson to get financial support from my wife (whom he had never met before) and some members of the Chicago-born Rosenwald family to buy from the widow of Colonel Knox the controlling interest in the *Chicago Daily News.* At that time we were barely acquainted. . . .
>
> Later Stevenson and I became much closer. We had long talks about the *Daily News,* and for the first time he made me well

acquainted with his indecisiveness. We met frequently. . . . I liked him. . . . There were a number of high calibre men I became intimate with in wartime Washington and I could make a rather long list of those whom I could not possibly think of as a suitable President. Adlai was one of them, and I just about told him so. [79]

Shortly before John F. Kennedy took office Ascoli wrote:

At the end of 1960, everything is so different that I feel somewhat nostalgic. We now have an administration . . . in which most of the men at the top are familiar with *The Reporter.* Some of them have written for us: first of all President Kennedy, a man who likes reading and thinking. He may even find the time to look at what we write. It's all a big change. During the last eight years, in commenting and reporting on the Eisenhower administration, we have leaned over backward in the effort to be helpful whenever we had a chance. But we were given few chances.

Among the subscribers who have been, as far as we know, satisfied with our work are a number of members of the cabinet like Robert S. McNamara, Dean Rusk, Stewart Udall. The danger that with so many friends and readers in high position we might get stuffy is, to say the least, remote. Our usefulness to the administration is in direct relation to our independence of it. [80]

The barb contained in this statement—"He may even find the time to look at what we write"—reflects Ascoli's basic problem: he did not like the man at the top. In 1971 he recalled, "John Kennedy and I had a couple of very *bad* long talks. We didn't like each other." He added, "He was not a very intellectual man. He didn't like ideas." The comments of one of the more perceptive of Ascoli's former staff members probably contain a nugget of truth:

In his relations with Ascoli, Kennedy may have been a victim of his famous staff work. His staff misread Ascoli, and therefore so did he. It's almost as if they had prepared Kennedy for his talks with Ascoli by programming a machine and then pressing a button marked "Jewish intellectual." It was the wrong button.

Ascoli commented in 1971 on his first talk with Kennedy, which took place on January 8, 1960:

> In describing the interview to some staff members . . . I said I had had a conversation with a talking Univac. Most of my questions, I thought, had hit a button, promptly followed by a recorded message, accompanied by a staggering amount of detail. Once or twice, I tried to figure out a question to which the answer might not have been recorded, and I had the feeling of something like a whirr. Then, there was a steely look in his eyes. . . .
>
> He brought the conversation to an unexpectedly rude end when he was approaching the door. "How many subscribers have you got?" I told him the figure which, at that time, must have been around 150,000. "Well," he said, "they are the hard core of those working for Humphrey and Stevenson." "Did I come out for either of them?" I answered. "Wasn't I more critical of them both than you?" He shrugged his shoulders and rushed out. . . .
>
> Incidentally, I heard that the story of the "Jewish intellectual" was circulating among the staff, but it did not make any sense to me.[81]

Despite this inauspicious beginning, however, Ascoli wrote an item for "The Reporter's Notes" a few months later deploring the raising of the "Catholic issue"—which he felt was a false one—by the news media, and Kennedy sent him a warm letter on May 20, 1960, saying that the item had been extremely helpful to him.

Part of the uneasiness Ascoli felt about Kennedy was due to Kennedy's youth. Shortly after Kennedy's nomination, Ascoli wrote: "Poor young man, poor rich man—he has to outgrow so many things. He has to outgrow wealth, family, platform, advisers."[82] After the Bay of Pigs he wrote: "Perhaps it was inevitable that a young President had to undergo his education in public, and prove his fitness to outgrow the blunders of inexperience."[83] *The Reporter* did formally endorse Kennedy in September 1960, but it surely must have been one of the more lukewarm endorsements he received: "We have a chance with Kennedy; we are sunk with Nixon."[84]

Ascoli was definitely not impressed with Kennedy's inaugural address, and said so in print. Among other things, he referred to one of the more celebrated passages—"Let us not negotiate out of fear, but let us not fear to negotiate"—as a "spiritual hiccough."[85] Ascoli soon learned that "the president was quite mad at me." He was invited by Kennedy to Washington on March 1, 1961, for luncheon and a second talk that "made the first one a love feast by comparison." As Ascoli recalls it:

> We had barely started the first pre-luncheon drink when he asked me, "Why didn't you like my inaugural address?" I soon realized that he had learned my comments practically by heart. . . . At the beginning I tried to defend myself by saying that I couldn't read anything, particularly an historical State document, without a critical eye. . . . I did not like that kind of compressed high-sounding oratory. It was mostly a condensation of ghost-writing and, anyway, great thoughts cannot be adequately expressed in a short speech. "Would you say the same about the Gettysburg Address?" he asked. Yes, I said. I got through it all at the expense of Lincoln, one of the few American Presidents whom I had adored well before coming to this country. At times, I had the feeling that I was a professor, talking to an arrogant and not too well prepared graduate student. I was praying God that the ordeal would come to an end, but Kennedy kept it going. What did I know about the Congo? A fantastic series of data was poured out at me. Was I sure Patrice Lumumba was a Communist? I haven't seen his card, I replied, but I knew enough about his record from some of my U.N. friends. Then he brought up Nkrumah. It went on for about two hours. At the end I was a wreck.[86]

Ascoli felt that he could not tell this story publicly and so remained far more critical of Kennedy privately than he was in his writing. He did not like his own attitude of criticism, and when he felt he could praise Kennedy, he did so generously. Commenting on the political risks Kennedy was taking in fighting for a Democratic majority in the Congressional election of 1962, he wrote: "He is a brave, a very brave man."[87] And his comments on Kennedy's television interview of December 17, 1962, were headed "A Great Performance."[88]

Still his doubts emerged in print. He was dubious about the "generation of middle-aged youngsters" who were "running around all the time" on the New Frontier.[82] And two youthful journalists, Worth Bingham and Ward S. Just, took a skeptical look at "All the Bright Young Men" for *The Reporter* (August 16, 1962). Ascoli was also bothered by Kennedy's excessive concern with his own image: "The chumminess between the chief leaders in the White House and influential purveyors of reflection and information beclouds the public view of the administration's doings."[90] But the title of the editorial in which he voiced this complaint, "A Few Anguished Questions," suggests that Ascoli was not fully happy about having to raise such issues.

Ascoli's dislike of Robert Kennedy was openly expressed in *The Reporter*, even before the younger Kennedy began speaking out against President Johnson's Vietnam policy. When he ran for the Senate in 1964, the magazine supported his opponent, Kenneth Keating. Ascoli disliked Robert Kennedy's youth, his "ruthlessness" and "opportunism," and perhaps most of all, his trading on the Kennedy name. When *The Reporter* closed, Ascoli said he was proud that it had opposed "any attempt to establish a Buonaparte dynasty."[91] In the issue of March 23, 1967, Ascoli wrote a strong and controversial editorial, "The Two USAs," depicting what he felt had happened to America since President Kennedy's assassination. He believed that it marked the beginning of a tragic split in the country. On one side were those who,

> even if they tried, could not bring themselves to recognize that the Constitution had given the country a new President. Even after the great electoral triumph of Lyndon Johnson, these people have become if anything more unreconciled to the presence of that usurper in the White House. The restoration must take place and a line of succession according to a new legitimacy must be made first into a tradition, and then into sacred law.

Robert Kennedy and his fellow-critics of President Johnson represented one U.S.A. "Do I need to tell *The Reporter*'s readers," asked Ascoli, "that I belong to the other USA? . . . I am proud to belong to that America which has its leader in Lyndon Johnson."

The Reporter had contained favorable articles on Johnson when he was the Senate majority leader in the 1950s; and as President, Johnson appealed to Ascoli because he combined liberalism at home with a certain toughness in foreign policy—like Truman. Ascoli liked strong leaders: "My last two heroes were Charles de Gaulle and Lyndon Johnson."[92] The fact that Johnson lacked Kennedy's "style" and "image" was in Ascoli's eyes only an advantage:

> There is no emotionalism in the relationship between Lyndon Johnson and the people. In the opinion of some, this gives a gray color to his image. If this means that there is distance felt on both sides between the President and the people, we cherish this distance. Only under Harry Truman had we a man in the White House who was not a source of complexes—fatherly or otherwise—throughout the nation.[93]

Ascoli did not care for many of Johnson's mannerisms, for some of his rhetoric, or for all of his programs; he was never an uncritical admirer. (In fact, the only two men in public life that Ascoli seems to have admired with no expressed reservations were those two very different personalities, Harry Truman and Dag Hammarskjöld.) Ascoli found Johnson's State of the Union message in 1966 "too long, crammed with too many subjects. . . . This man is not a visionary or a radical: he is an exuberant middle-of-the-road extremist."[94] Later that year he chided both the anti-Johnson intellectuals for their lack of restraint and Johnson himself for refusing to listen to criticism. Johnson, he felt, suffered from being "the one-man embodiment of this country, with all its immense vigor and ever-deluded ambitions—including the ambition to be universally loved."[95]

The growing opposition to Johnson among many intellectuals only strengthened Ascoli's determination to support the President. The conflict between the intellectuals and Johnson was in part a personal one and in part a cultural one, and Vietnam may have been the occasion of the conflict rather than its basic cause.[96] The paranoid McCarthyism of the right, which had infected the early 1950s, had been replaced by its mirror-image, a paranoid McCarthyism of the left; both McCarthyisms were characterized by the belief "that whatever has gone wrong was *planned* to go wrong."[97] It is only logical that a man who had fought the old McCarthyism should vigorously oppose the new variety. Shortly after announcing the closing of *The Reporter* Ascoli said defiantly: "I don't give a hoot in hell about the liberal community. The thing that makes me proudest about this magazine is its defense of the Johnson Administration when everyone else was going against it."[98]

And yet Ascoli was writing at the same time, with regard to the President's address of March 31, 1968, that he did not expect "that the President would run out on his pledge to the people of South Vietnam, and at the same time run out on the American electorate by proclaiming himself a lame duck." He added sadly that "for those of us . . . who have maintained confidence in the President, sometimes at a rather heavy cost, the sense of having been let down is hard."[99]

A month later Ascoli wrote that by withdrawing from the presidential race Johnson "destroyed any advantage we might have found in negotiating":

> Lyndon Johnson has thus regained most of his consensus, and lost most of his power. I wish him well, as is the due of a man who has overspent himself for his country. I am proud to have been for him, and I wish I could still be. But the chapter is closed.[100]

And in November 1968, still disillusioned with Johnson's actions, Ascoli voted against him, in effect, by casting his ballot for a man he and his magazine had criticized steadily for nearly twenty years, Richard M. Nixon.

Thus, underlying *The Reporter*'s commitment to many liberal domestic causes over nineteen years and affecting its relations with four administrations was one basic theme: America's relationship with the rest of the world. At *The Reporter*'s end, as at its beginning, this remained the crucial issue.

NOTES

1. Ascoli, "*The Reporter* in 1953," *The Reporter*, December 23, 1952, p. 7.
2. Bruce Bliven, *Five Million Words Later*, (New York: John Day, 1970), p. 202.
3. Ibid., pp. 268-72.
4. "The *Nation* Censors a Letter of Criticism," *New Leader*, March 19, 1951, p. 17. Cf. "Soul Searching on the Left," *Time*, April 2, 1951, pp. 44-45.
5. "Exit from the *Nation*," *Time*, May 21, 1951, p. 51.
6. Editors of the *Partisan Review*, "Liberalism, Libel, and André Gide," *Partisan Review* 18 (May-June 1951), 365-68.
7. Granville Hicks, "The Liberals Who Haven't Learned," *Commentary*, April 1951, pp. 319-29.
8. Ascoli, "Station Identification," *The Reporter*, September 26, 1950, p. 1. The desire to dissociate *The Reporter* from the established liberal magazines did not deter Ascoli from taking part in a forum sponsored by the Nation Associates in the spring of 1950. His remarks there were published in the May 20, 1950, issue of the *Nation* (pp. 489-90) under the title, "What Total Diplomacy Means."
9. Ascoli, "What We Stand For," *The Reporter*, December 20, 1949, p. 2.
10. "Wordsmith at Work," *Newsweek*, July 12, 1965, p. 27.

11. Interview with Horton, January 1971.

12. Letter from Ascoli to the author, May 7, 1971.

13. Interview with Bendiner, January 1971, and letter from Ascoli to the author, May 17, 1971.

14. "Sorry, Senator," *The Reporter*, October 24, 1950, p. 1.

15. "Use of Magazine by Diplomats Hit," *New York Times*, October 8, 1950, p. 19. See also Ascoli's reply, *New York Times*, October 9, 1950, p. 43.

16. *Congressional Record*, 81st Cong., 1st sess., 95, pt. 10: 13269 (September 26, 1949); 82nd Cong., 2nd Sess., 98, pt. 5: 6758-59 (June 6, 1952).

17. *Congressional Record*, 82nd Cong., 2nd Sess., 98, pt. 5: 6759 (June 6, 1952).

18. *Congressional Record*, 83rd Cong., 2nd sess., 100, pt. 12: 15182-90 (August 19, 1954). Ascoli replied in detail to Bridges' charges in a letter to Senator Herbert Lehman, which was inserted in the *Congressional Record*, 83rd Cong., 2nd sess., 100, Appendix: A6583-85 (September 3, 1954). See *The Reporter*, September 23, 1954, pp. 2-4.

19. *New York Post*, February 5, 1954.

20. *Foreign Service Journal*, June 1954, p. 4.

21. *New York Times*, January 14, 1962, p. 42

22. "Not for El Segundo," *The Reporter*, March 29, 1962, p. 17.

23. Robert Lekachman, "Political Magazines," *Book Week*, November 17, 1963, pp. 24, 28.

24. Robert Sherrill, "Weeklies and Weaklies," *Antioch Review*, Spring 1969, p. 31.

25. Alastair Buchan, "Witch-Hunting," *London Observer*, June 25, 1953.

26. Willmoore Kendall, "Dreams of Togetherness. . . ," *National Review*, July 25, 1956, p. 15.

27. Interview with Irving Kristol, January 1971. Ascoli, characteristically, disagrees with this label as applied to himself.

28. "Liberal Anti-Communism Revisited," *Commentary*, September 1967, p. 67.

29. Ascoli's office files contain a mimeographed release put out by the UAW-AFL "Labor Great Books Program" in Milwaukee (April 24, 1950) which says: "*The Reporter* is highly recommended to all local unions and readers of the LGBP releases."

30. *The Reporter*, November 1, 1956, p. 6.

31. Cf. Andrew Knight, "America's Frozen Liberals," *Progressive*, February 1969, p. 34: "The liberal believed in the efficacy of central government until, to his surprise, he found both Senators Kennedy and McCarthy arguing (with the conservatives) last year for a devolution of power"; *Newsweek*, November 20, 1967, p. 60: "Contrary to the cherished belief of the old liberalism, the Federal bureaucracy is not the ideal instrument for handling social issues which must ultimately be settled at the local level"; and Theodore H. White, *The Making of the President—1968* (New York: Atheneum, 1969), pp. 196-97.

32. Ascoli, "What Price Big Government?" *The Reporter*, March 14, 1950, p. 5. E. B. White was expressing similar ideas at about this time in the *New Yorker*. See *Letters of E. B. White*, ed. Dorothy Lobrano Guth (New York: Harper and Row, 1976), pp. 376-77.

33. Ascoli, "The Strain on Our Civil Liberties," *The Reporter*, August 30, 1949, p. 3.

34. Ascoli says: "I owe to that issue at least one friend, and one enemy. The friend was Harry Truman." (Letter to the author, April 23, 1971). The issue drew appreciative letters from Harry Truman's aide, Charles G. Ross, on behalf of the President, and from Herbert Lehman, Jacob Javits, and Norman Cousins, among others. (*The Reporter*, July 4, 1950, pp. 2-3.) *Time*'s cover story on McCarthy—a strong attack—did not come until October 1951, more than a year later.

35. "McCarthy's Blunderbuss," *Nation*, March 18, 1950, pp. 243-44; "McCarthy's Great Red Scare," *New Republic*, March 20, 1950, pp. 5-7; Amy Loveman, "Diversity and Union," *Saturday Review*, April 1, 1950, pp. 22-23; Richard Rovere, "Letter from Washington," *New Yorker*, May 13, 1950, pp. 96-103; "Ideas Can Be Dangerous," *Time*, April 17, 1950, p. 29.

36. John Hoving, "My Friend McCarthy," *The Reporter*, April 25, 1950, pp. 28-31.

37. Ascoli, "Trying Times," *The Reporter*, June 17, 1965, p. 14.

38. Cater's early analysis of McCarthy's use of the press anticipates the observations made later in Richard Rovere, *Senator Joe McCarthy* (New York: Harcourt Brace, 1959), pp. 166-73, and has become a standard view among historians of journalism. See also Cater, *The Fourth Branch of Government* (New York: Vintage, 1959), pp. 68-74.

39. See his letter to the editor, *The Reporter*, July 4, 1950, p. 2.

40. *The Reporter*, November 11, 1952, p. 3.

41. *The Reporter*, November 18, 1954, pp. 8-9.

42. *The Reporter*, February 6, 1958, p. 8.

43. *The Reporter*, July 10, 1958, p. 8.

44. *The Reporter*, December 25, 1958, p. 8.

45. The *New Republic* had run a special twenty-page section, "The Negro: His Future in America," in its issue of October 18, 1943; *Newsweek*'s special issue, "The Negro in America," appeared in November 1967.

46. *The Reporter*, January 31, 1950, p. 2.

47. E.g., "Slums, Ghettos, and the G.O.P.'s 'Remedy' " (May 11, 1954) and "Segregation, Housing and the Horne Case" (October 6, 1955), both by Charles Abrams, New York State's rent administrator; "The Army and Its Negro Soldiers" (December 30, 1954) by Leo Bogart; "Washington: 'A Model for the Rest of the Nation' " (December 30, 1954) by Douglass Cater; and "A Theologian's Comments on the Negro in America" (November 29, 1956) by Reinhold Niebuhr.

48. Ascoli, "The Courage of Prudence," *The Reporter*, March 8, 1956, p. 12.

49. *The Reporter*, April 5, 1956, pp. 6-7.

50. *The Reporter*, May 10, 1962, pp. 6-8.

51. The *New York Times*'s five-inch story was headlined "Faulkner Believes South Would Fight" (March 17, 1956, p. 17). *Newsweek* wrote, "William Faulkner of Mississippi supplied some frightening copy to the magazine *Reporter* last week" and quoted a sentence on Southern resistance (March 26, 1956, p. 90). *Time* used two of Faulkner's sentences on Southern resistance as part of a cover story on the South (March 26, 1956, p. 26). Faulkner's letters of protest appeared in *Newsweek*,

April 9, 1956, p. 58; *Time*, April 23, 1956, p. 58; and *The Reporter*, April 19, 1956, p. 7.

52. Ascoli, "To Our Negro Brothers," *The Reporter*, August 10, 1967, p. 12.
53. Ascoli, "The Segregated Ballot Box," *The Reporter*, June 27, 1957, p. 8.
54. Ascoli, "The Courage of Prudence," *The Reporter*, March 8, 1956, p. 12.
55. Ascoli, "To Our Negro Brothers," *The Reporter*, August 10, 1967, p. 12.
56. *The Reporter*, November 17, 1966, p. 4.
57. Ascoli, "Central High and Quemoy," *The Reporter*, September 18, 1958, p. 12.
58. Ascoli, "The Birth of a Nation," *The Reporter*, June 6, 1963, p. 12.
59. Ascoli, "Unbelting the Earth," *The Reporter*, May 24, 1962, p. 10.
60. Ascoli, "The Peace Games," *The Reporter*, September 12, 1963, p. 26.
61. "Our Liberalism," *The Reporter*, March 28, 1950, p. 1.
62. "Station Identification," *The Reporter*, September 26, 1950, p. 1.
63. Eric Sevareid wrote in response to this editorial, "I'm a bit nervous about the rush to play co-God with Russia, too" (*The Reporter*, June 4, 1964, p. 6).
64. Ascoli, "The Revolt Against Freedom," *The Reporter*, February 8, 1968, p. 13.
65. Ascoli, "Farewell to Our Readers," *The Reporter*, June 13, 1968, p. 18.
66. Ascoli, "The Cool Precision of J.F.K.," *The Reporter*, February 16, 1961, p. 22.
67. *The Reporter*, December 29, 1955, p. 4.
68. *New York Times*, September 14, 1952, p. 80.
69. *New York Times*, October 11, 1952, p. 18.
70. Ascoli, "The Miracle," *The Reporter*, August 11, 1955, p. 13.
71. Letter from Ascoli to the author, May 7, 1971.
72. Ascoli, "And They Say We Are Doing Fine," *The Reporter*, April 5, 1956, p. 10.
73. Ascoli, "Back to What 'Normalcy'?" *The Reporter*, October 27, 1953, p. 9.
74. Ascoli, "The Miracle," *The Reporter*, August 11, 1955, p. 13.
75. Ascoli, "The Reporter's Notes," *The Reporter*, February 2, 1961, p. 12.
76. *The Reporter*, October 14, 1952, p. 1.
77. Her attack was so effectively worded that Victor Lasky lifted a number of phrases from it and used them verbatim to condemn the prose of John F. Kennedy. See Rivers, *The Opinionmakers* (Boston: Beacon Press, 1965; reprinted 1967), pp. 48-49.
78. "The Waiting Game: 'The Reporter' and Adlai Stevenson," *New Republic*, January 16, 1956, pp. 7-8.
79. Letters from Ascoli to the author, May 7 and May 10, 1971.
80. Ascoli, "A Report to Our Readers," *The Reporter*, January 5, 1961, p. 14.
81. Letters from Ascoli to the author, May 7 and May 10, 1971.
82. Ascoli, "From the Sports Arena," *The Reporter*, August 4, 1960, p. 15.
83. Ascoli, "Foreign Policy After Cuba," *The Reporter*, May 25, 1961, p. 18.
84. Ascoli, "The Only Choice," *The Reporter*, September 29, 1960, p. 14.
85. Ascoli, "The Inaugural Address," *The Reporter*, February 2, 1961, p. 10.
86. Letter from Ascoli to the author, May 7, 1971.

87. Ascoli, "The President's Big Risk," *The Reporter*, October 25, 1962, p. 14.

88. *The Reporter*, January 3, 1963, p. 12.

89. Ascoli, "All Roads Lead to Washington: Notes from the National Capital," *The Reporter*, February 1, 1962; "Shifting Priorities," *The Reporter*, July 5, 1962, p. 8.

90. Ascoli, "A Few Anguished Questions," *The Reporter*, April 25, 1963, p. 22. A year earlier (April 12, 1962) *The Reporter* had run an article by Worth Bingham and Ward Just on this subject, "The President and the Press."

91. Ascoli, "Farewell to Our Readers," *The Reporter*, June 13, 1968, p. 18.

92. Ibid.

93. Ascoli, "The President-Candidate," *The Reporter*, September 10, 1964, p. 22.

94. Ascoli, "The President's Speech," *The Reporter*, January 27, 1966, p. 22.

95. Ascoli, "Uncle Ezra's Legacy," *The Reporter*, November 3, 1966, p. 16.

96. See Goldman, *The Tragedy of Lyndon Johnson* (New York: Knopf, 1969), pp. 418-75, and Irving Kristol, "American Intellectuals and Foreign Policy," *Foreign Affairs*, July 1967, pp. 594-609.

97. Garry Wills, *Nixon Agonistes* (Boston: Houghton Mifflin, 1970), p. 60. Cf. John Roche, "Dissent, Consensus, and McCarthyism," *The Reporter*, December 12, 1965, p. 10.

98. "Reporter's End," *Newsweek*, April 22, 1968, p. 67.

99. Ascoli, "The Presidency," *The Reporter*, April 18, 1968, p. 14.

100. Ascoli, "On Lyndon Johnson," *The Reporter*, May 16, 1968, p. 8.

Chapter Five

The Crucial Issue: America and the World

The American national interest is definitely tied to the welfare of a large section of the world, but at the same time America is not rich enough or powerful enough to endow the world or rule it.

Prospectus for *The Reporter*, 1949

Many of my liberal friends have wondered how The Reporter *can take such a liberal stand on domestic issues and at the same time such a firm stand on the war in Vietnam. What these people fail to grasp is that you have stood consistently on the side of freedom for all mankind from tyranny in whatever form it should appear—be it Communism, fascism, or racism—and for freedom of the individual from the abuse of power, whether wielded by a rednecked sheriff, a corrupt politician, a union boss, or an ambitious U.S. Senator.*

A subscriber in Michigan, 1968

"The basic idea of *The Reporter,*" Ascoli wrote in 1954, "is that this nation of ours is in the business of world leadership for keeps."[1] He phrased this idea in many ways, but throughout the nineteen years of the magazine's existence it was always clear that the fundamental concerns of *The Reporter* were global in scope. A former staff member recalls suggesting to Ascoli once in the 1950s that the magazine run a piece on the U. S.

economy. Ascoli wanted to know why that subject was so important just then.

"Well, Max, the jobless rate has gone way up, and people are worried. It's that simple."

"Worried about the jobless rate?" muttered Ascoli. "I'm worried about the planet!"

Ascoli's concern about the planet appeared to coincide generally with the concerns of the liberal community during most of the 1950s. By the end of that decade, however, certain divergences became noticeable, and by the mid-1960s they were conspicuous. Some of Ascoli's readers—and staff members—began to react to his editorials on foreign policy and to accuse him of turning conservative. But the two basic principles expressed in his foreign-policy editorials—his insistence on the need for American involvement in world affairs and his reiterated belief that freedom and Communism were philosophically and ethically irreconcilable—remained constant. Even if the world—or perhaps only his critics' view of the world—was changing, he could maintain that he was consistent. As to whether his views could consistently be labeled "liberal," that depends on how one defines a liberal foreign policy.

CHARACTERISTICS OF A LIBERAL FOREIGN POLICY

Defining a liberal foreign policy is not as easy as it might seem. Two of the contributors to a book of foreign-policy essays that appeared in the 1960s, *The Liberal Papers*, suggest some of the difficulties of defining either a defense policy or a foreign policy that can be confidently labeled "liberal":

> It is difficult to speak in terms of a specifically "liberal" defense policy. . . . It is useless, for example, simply to inveigh against current tendencies toward the militarization of our policy and economy if these tendencies are no more than symptoms of deeper historic causes to which liberals may have contributed as much as conservatives or those of any other persuasion.

The writer concludes that in constructing a "liberal" military policy "the moderate course, even if illogical[!], is likely to turn out best in the end."[2]

As for foreign policy,

> the liberal is beset by particular difficulties because he believes, on the one hand, in the independence of nations and the self-determination of peoples, and, on the other hand, in respect for human rights based upon the values of individual freedom, democracy, constitutionalism, and social progress. He is, therefore, at the same time, an advocate of peaceful coexistence and mutual toleration. . .; and a militant crusader for the American interpretation of freedom, democracy, constitutionalism, and progress. Emphasis upon the first would seem to tolerate tyranny, regimentation, and oppression . . . in many nations, while the second would seem to require non-recognitions, propagandas, and interventions, maintaining high international tensions and cold war, threatening hot war likely to destroy mankind. The dilemma is not unlike that faced by Lincoln.[3]

As often with domestic policy, then, so with foreign policy the liberal tends to be the man in the middle. And because he is aware of the perils of extremes, he realizes that the essential elements in foreign policy are normally matters of balance, timing, nuance, and semantics. As conditions in the world change, the emphasis of foreign policy needs to change. The changes can be very subtle, and popular labels like "hawk" or "dove" are often overly simplistic.

In detecting new conditions in the world, however, it is not always easy to maintain perspective. This is particularly true with regard to changes in the Communist world. Some observers seize eagerly on every sign of liberalization in Communist nations, overlooking the repression that still exists; others focus grimly on all signs of continuing repression and overlook the positive changes that have occurred. Those who are particularly aware of America's shortcomings are sometimes disposed, by way of reaction, to be relatively tolerant of the shortcomings of Communist nations. And those who speak most strongly about the evils of Communism may seem strangely blind to America's failings. A lopsided conscience can sometimes develop: in the controversy over Vietnam, one group of Americans raged against the American use of napalm; another group fulminated with equal passion and moral fervor against the terrorism practiced by the Viet Cong. Life is full of bear traps, but they seem to be especially thick in the deep woods of foreign policy.

A democracy, as Alexis de Tocqueville pointed out in a famous passage, has great disadvantages in conducting foreign affairs, for it "can only with great difficulty regulate the details of an important undertaking, persevere in a fixed design, and work out its execution in spite of serious obstacles. It cannot combine its measures with secrecy or await their consequences with patience."[4] Thus Americans can get tired of long-drawn-out foreign enterprises, and then the latent strain of isolationism in American thought may begin to become dominant.[5]

One sees this in the disillusionment among liberal intellectuals that followed World War I. In what seems an uncannily prophetic strain, the *Nation* editorialized, in its issue of November 3, 1920: "Whenever liberalism strikes hands with war it inevitably goes down. . . . For war and liberalism to lie down together anywhere, at any time, with any excuse, means only one thing—disaster to liberalism."[6] The disillusionment of the 1920s led the disenchanted liberals of the 1930s "to distort their moral perceptions beyond all reason by concentrating upon the defects of the democratic powers and dwelling upon the evils of the past, while seeking extenuating circumstances for Fascist conduct and minimizing the present danger." They developed the comforting theory that the danger of Fascism in terms of military aggression was superficial; the real danger was economic and social maladjustment. Thus domestic reform could serve as a substitute for diplomacy.[7]

This kind of disillusionment and isolationism did not immediately follow World War II. Although some conservatives wanted America to withdraw from the world, liberals tended to be internationalist, and they gave strong support to the United Nations. A liberal President—with the general support of the liberal community—issued the Truman Doctrine, initiated the Marshall Plan, ordered the Berlin Airlift in response to the Soviet blockade, acted decisively in Korea, and made the initial commitment of aid to Indo-China. But when this last decision led, more than a decade later, to American combat involvement in an inconclusive land war in a part of Asia where America's self-interest was not apparent to many American citizens, a widespread disillusionment and isolationism among liberals reasserted itself. Liberals—who often considered themselves to have been "radicalized" by a combination of the civil-rights movement and the Vietnam War—began to insist that America's domestic problems should have first priority.[8]

THE REPORTER'S FOREIGN POLICY

The basic elements in *The Reporter*'s position on foreign policy—opposition to isolationism, Fascism, and Communism, and advocacy of western unity and of agreements with the Soviet Union on arms control—sound like tenets held by a large part of the liberal community in the 1950s. But a distinctive pattern of philosophical and personal beliefs lay behind Ascoli's advocacy of them, and when some of these beliefs began to become apparent by the beginning of the 1960s, a divergence between *The Reporter* and certain elements of the liberal community became evident.

Ascoli's basic value, as he made clear repeatedly, was freedom, although he was not always any more successful than other liberals in defining that abstraction. But he was clear about what he considered to be its three most dangerous enemies—isolationism, Fascism, and Communism. And he saw the three as so intimately interrelated that to give him a single label—"internationalist," "antifascist," or "anticommunist"—is to oversimplify and misrepresent his views. He continually expressed his ideas in *The Reporter*, but because he was often addressing himself to a current question and writing against the pressure of a deadline, his editorials have at times an enigmatic quality; they express his beliefs in a truncated and unsatisfactory form.

One of the most complete early expressions in *The Reporter* of his views on freedom and its enemies can be found in an article he wrote when the magazine was just one year old, "Our Political D.P.'s" (April 25, 1950). The article was written in response to an article in the same issue, "What Can Ex-Communists Do?" by ex-Communist Isaac Deutscher, because as Ascoli says in his opening sentence, "there are assertions in Isaac Deutscher's article that cannot be left unchallenged." Deutscher's article maintains that the ex-Communist frequently remains a sectarian, "an inverted Stalinist." He argues that "the only dignified attitude the ex-Communist can take is to rise *au-dessus de la melée*" like some of those who had been disillusioned with the French Revolution.

In reply, Ascoli points out the weakness of arguing by historical analogy and argues that ex-Communists simply cannot keep aloof from the conflict between freedom and Communism. He concludes: "Perhaps it is good to have them in our midst, even if some of the ex-Communists are unbearably irritating. They make us realize how shallow and unappealing our values have become." (Ascoli's disagreement here with Deutscher seems to have been largely a matter of emphasis, since Ascoli never had much use for the widely publicized ex-Communists of the 1950s.)[9]

Most of the former Communists, Ascoli notes, "wanted to fight for the underprivileged or against fascism. Disgust with some of the soft-headed and ineffectual practitioners of freedom called liberals drove many ardent and unselfish men and women into Communism." Their difficulty was that they did not know the meaning of freedom; so he defines it:

> This is what freedom means: the right to go back to oneself and judge what one has done, or what one has been made to do, so as to gain the experience one needs to do better. Freedom means the constant growth and enrichment of the individual, and through him, of the society in which he lives.

Opposed to freedom is what he calls the "problem of our times":

> Ours is the problem of the human person who has been trapped by too many technological or political machines and is in danger of being extinguished. It is the problem of a civilization threatening to turn into an unmanageable force of nature.

Because Communism, for Ascoli, represents the culmination of this process, he calls it "this mercilessly organized soullessness." This is hardly the thinking or language that characterizes much of the standard American anticommunist rhetoric. Rather, it reflects the deeply felt ethical and spiritual values that characterize the indictment made of Communism by Soviet dissident André Sinyavsky or the indictment of the technological society by the French social philosopher Jacques Ellul.[10]

Thus, Ascoli continues, he does not want the battle for freedom to "degenerate into an anti-Communist crusade," which it will "if we take to our side as full partners all the regimes and interests that Communism has threatened." And yet America does need "allies and partners with whom to share the responsibility of its power," for it is "engaged in a world civil war unprecedented in character and scope. The enemy is increasingly becoming the embodiment of everything that was wrong with our political and economic system." A few months later, Ascoli insisted that *The Reporter* was not going "to be mesmerized by the Communist danger, or to waste its energies psychoanalyzing the Communist mind. While fighting Communism it tries to look beyond it—to the world we are going to live in when the countries infected by the Communist poison are finally reabsorbed into our civil society."[11] Ascoli's expression of his antifascist beliefs, a decade earlier, is strikingly similar. Shortly before the United States entered World War II, Ascoli insisted that the war was really "the civil war of a world that modern civilization and modern technology have unified but that the leading democracies have failed to unite. . . . It is not exactly Hitler alone that we must fight, or a peace after Hitler that we must prepare, but we must fight a mode of living and define our life aims."[12]

Ascoli regarded Fascism and Communism as equally inimical to freedom:

> A centralized, planned society, in which the government acquires the power to decide on labor conditions and on prices and on profits, is conceivable only if it is a totalitarian one—and it doesn't make much difference whether the totalitarianism is of the Communist or of the Fascist variety.[13]

And yet he felt one could and should distinguish between them:

> There is a fatuous and mean quality in fascist totalitarianism that distinguishes it entirely from Communist totalitarianism. The Communists sacrifice the lives, the rights, and the happiness of human beings in order to verify a lunatic conception of history. Fascism, having no conception of its own, glibly adopts or repudiates any number of ideologies concocted by its hired intellectuals. . . .
>
> Communism is hell endured for the sake of a sham heaven. Fascism is hell for the hell of it.[14]

On the practical level, Ascoli distinguished between the two in seeing Communism as primarily an external danger to America and Fascism as primarily an internal danger. The greatest danger of Fascism, he felt, came from isolationism, and this is why he always opposed any isolationist tendencies in America—whether stemming from fear, despair, laziness, self-centeredness, or a McCarthyite variety of anticommunism:

> We have fascism whenever a nation tries to find parochial, strictly nationalistic solutions to world-wide problems—like resisting Communism, striking a balance between capital and labor, establishing a national economy entirely independent of that of other nations. When a country attempts to segregate itself from the rest of the world and run itself as if it were the world, then we have fascism.[15]

Ascoli's perception of the danger from Communism, like that of many other liberals, was sharpened by events of the first postwar years. In 1944 he cautiously believed that coalitions between liberals and Communists would be possible in the governments of liberated Europe—provided that "liberals all over the world have an equally alert sense of international solidarity" as do the Communists.[16] By 1953 it was clear to him that he had been overly optimistic: "One thing is sure: There can be no possible sharing of responsibility between Communist and democratic parties within the government of a country. That has been tried all over Europe, east and west; invariably it has ended in disaster."[17]

On the other hand, coexistence between Communist and democratic governments, though frustrating at times, could pay off: "If we do all that is in our power to create conditions that actually make for freedom and for solidarity among free nations, then we have nothing to fear from co-existence, for we know where it will lead"—to a strengthening of the coalition of free peoples and a weakening of Communist tyranny.[18] He added in the same editorial, characteristically, that Eisenhower should be able to coexist with Malenkov; he had, after all, managed to go through the presidential campaign of 1952 coexisting with Joe McCarthy. Coexistence did not mean simply reacting to Soviet initiatives, on the one hand,[19] or expecting some utopian scheme that would be "that most terrifying of fool's paradises—a foolproof system for maintaining peace."[20] Rather, as he put it a few years later:

> Peace for our times can only be based on a series of temporary, limited, patchwork settlements, until some reduction of armaments can be achieved—and the reduction of armaments, too, must be a series of limited, patchwork agreements.[21]

Our "normal relations with the Communists," he wrote after the Cuban missile crisis, "should never be of war and can never be of peace. As long as they are Communists, they cannot possibly deviate from the ultimate aim of global conquest."[22]

ALLIED UNITY AND THE UNITED NATIONS

Ascoli believed that the most important element in combating Communism was the unity of the noncommunist world, and he felt that America needed to assert leadership in building this unity.[23] He believed that America should lead the noncommunist world not because it represented the *pattern* of government that others should slavishly imitate but because it symbolized the *potential* of human development:

> We are not the chosen people destined to save the world, and our Congress can never act as if it were the Parliament of Man. But somehow we feel that what has happened to men of many races who came here and became Americans should happen, *in many different and still unfathomable* ways, to the rest of mankind. We do not want to remake the world in our own image, for otherwise it would never be what we want it to be: a world of free men—really a free world, without quotes.[24]

American foreign policy, Ascoli believed, should show "inventiveness and creation" through the "building of the inter-Allied commonwealth, or rather of a network of interlocking commonwealths, all centered around the United States."[25] The creation of these interlocking commonwealths would surpass both nationalism and imperialism and give the West the ability to take the initiative in the reduction of armaments and in other aspects of competition with the Communist world.[26]

The practical importance of Allied unity as a bulwark against Communism in Europe had been the rationale for the creation of NATO in 1949 and the attempts to create the European Defense Community (EDC) in the 1950s and the Multi-Lateral Force (MLF) in the 1960s. Ascoli wrote relatively little about such organizational approaches to Western unity. For the importance of such unity, in his view, went far beyond any merely strategic or organizational considerations. Western unity symbolized the unity of civilization itself:

> The United States cannot isolate itself from the rest of the Western community without turning into a purposeless oversized thing—indefensible both morally and militarily.
>
> What makes the West so tightly knit a community that no member can leave it except by suicide lies in the fact that there is no alternative to the West or to its material and spiritual creativeness. What is called the West has no antithesis in its Communist imitation, or in a mythical East. The West is not a form of civilization, contemporary with other forms of civilization. The West *is* civilization.[27]

Ascoli's concept of the West—which he did not necessarily regard as just a geographical entity—led him to a persistent skepticism about the role the United Nations should play in United States foreign policy. He insisted:

> The best we can do for the U.N. and at the U.N. is to be ourselves: the greatest power and the leader of the West. It is there that in closest possible cooperation with our allies we must act according to those western interests and concepts which bind us to so many nations, some of which—like Japan, for instance—are far removed from what is geographically called the West.[28]

Therefore he disagreed with Adlai Stevenson's description of the United Nations, in the first year of the Kennedy Administration, as "the center and principal forum of our foreign relations." Rather, the "Western community and *not* the U. N. must be the center and principal forum of our national policy." He insisted that there was "no better way of reducing the U. N. to an unmanageable state than to follow indiscriminately the principle of 'back the U. N.' "[29]

Ascoli's skepticism about the United Nations and his outspoken opposition to Communism might be seen in retrospect as early signs of his divergence from important elements in the liberal community. But they were not so regarded at the time. For one thing, other prominent liberals shared some of his doubts about the United Nations. Senator Fulbright picked up his concept of interlocking commonwealths and used it as the basis for an article in *Foreign Affairs* in 1961, an article that *U.S. News*, a publication not notably in sympathy with either Fulbright or the United Nations, reprinted under the provocative title " 'U. N. Has Fallen Short; We Must Look Elsewhere' " (October 2, 1961).[30] And Ascoli's anticommunism was not conspicuous in the 1950s because he did not fit the stereotype of the anticommunist during that decade. He was an early, outspoken, and persistent opponent of Joe McCarthy; he had suggested the recognition of Communist China; he spoke out strongly against Franco and Peron and was extremely skeptical about Chiang Kai-shek; and he was a constant critic of John Foster Dulles.

Ascoli's delight in castigating Dulles, of course, seemed to place *The Reporter* squarely in the same ideological camp as the *Nation* and *New Republic*, which attacked the Secretary of State in such editorials and articles as "Dulles Should Go," "Secretary Dulles Must Go Now," "The Opacity of Mr. Dulles," "Let's Trade Dulles to China," and "Dulles Is Wrong."[31] Ascoli, who wrote in 1957, "I like to give occasional evidence of editorial forbearance by showing how, in spite of evermounting provocation, I can manage, for a couple of issues, not to berate Secretary Dulles,"[32] termed him a "national liability."[33] After the 1956 election, he wrote an editorial urging Eisenhower to replace Dulles with Adlai Stevenson.[34]

Ascoli's criticism of Dulles was not always consistent. In 1956 he called him "a man endowed with unfettered freedom from consistency."[35] Three years later he complained, "The perpetual motion of his body and of his mind succeeded in hiding the fixity of his position."[36] A few months before Dulles became Secretary of State and began to attack "neutralism," Ascoli himself had attacked neutralism in Europe as "nothing but naked pro-Communism."[37] And in 1955 Ascoli grudgingly praised Dulles' "new approach toward bilateral collective security" because it promised, if consistently followed through, to "dispose once and for all of the latest and most dangerous Communist fraud: neutrality."[38] The cause of Ascoli's animus toward Dulles seems to have been not so much the latter's ideas—the two men shared a strong spiritual antipathy toward Communism—as the way Dulles continually seemed to debase those ideas. Shortly before Dulles' retirement Ascoli tried to evaluate that "self-centered, overly energetic intellectual" and wrote that "few men in our times have contributed more to debase religious values by translating them into slogans." Describing Dulles as "exasperatingly devious," Ascoli wrote that "his mastery of language and of the rational process has led him to indulge in the coining of dazzling phrases that played the role of new policies, alarmed friends or foes, and expressed nothing."[39]

Ascoli liked the man whom Kennedy had chosen as his Secretary of State, Dean Rusk. In the spring of 1961 he described Rusk as "a good and wise man, who can prove his worth when the President gives him the power that the office of the Secretary of State demands"; in 1965 he called him "a man of superior knowledge and wisdom" and the following year characterized him as a "meticulously responsible, self-effacing public servant."[40] "The least that can be said about Dean Rusk," Ascoli wrote on this last occasion, "is that the enemies he has made are a tribute to him."

Many of these enemies were *The Reporter*'s enemies as well, for, like Rusk, *The Reporter* had come to be seriously out of phase with many liberals on foreign policy. During its first ten years *The Reporter* was in general agreement with many liberals on foreign policy. Then, during a watershed period from late 1959 to late 1964, the divergence of *The Reporter* from many liberals over the issue of Communism became increasingly evident. Finally, from 1965 until the closing of the magazine in 1968, *The Reporter* was definitely at odds with many liberals over Vietnam.

In the early 1950s one major issue was China, and *The Reporter* had incurred the wrath of conservatives by suggesting, in its issue of January 3, 1950, the recognition of Communist China. The suggestion was accompanied by careful qualifications: recognition of a government should not be taken as approval of that government; Communist China was a fact, and that fact should be recognized. And firmness was needed: "Decent diplomatic relations can be established—when the Red Leaders understand that the establishment of these relations can be only on a strict 'no-nonsense' basis."[41] Ascoli had little use for Chiang's government, as the China Lobby series made clear in the spring of 1952. Yet his real concern was not Chiang's shortcomings but the future of the inhabitants of Taiwan. In 1951 he urged that they should be allowed to determine their own destiny, through a plebiscite conducted under United Nations supervision. Several years later he advocated a United Nations trusteeship for Taiwan as a condition for the admission of Communist China to the United Nations.[42]

In urging this solution to the Taiwan question, Ascoli may have been overly optimistic about the ability of the United Nations to extricate the United States from a sticky situation. But in connection with a second major foreign policy issue of the 1950s, the Suez crisis, he criticized the Eisenhower Administration for a similar reliance on the United Nations. He felt the administration was jeopardizing the United Nations by using it as a shield for American indecisiveness. "Our government foredooms Dag Hammarskjöld to failure or ridicule," he wrote, "by attributing to him unlimited capacity to fix universal troubles, and it smothers the U.N. by massive reliance on it."[43] Ascoli's condemnation of such "peacemongering"—it was one of the five editorial positions that he named with pride in the tenth anniversary issue of *The Reporter*—brought some accusations that he had "turned against the U. N."[44] He replied that he was merely recognizing the existence of power blocs in the United Nations and was advocating a policy that would save the organization: "Indeed, it can be said that the best service our country can render to itself and to the U.N. is to re-establish the cohesion and the unity of our own bloc."[45] He felt that it was important to try to distinguish between the United Nations' function as a symbol of human hopes and what he believed to be its reality as "a many-layered center of international intrigues and of countless, unrelenting efforts to relieve human suffering; an object of starry-eyed gushiness and of cheap cynicism; the place where men and women utterly dedicated to goodness meet Krishna Menon."[46]

Nearly five years later, just after Dag Hammarskjöld's death, Ascoli finally revealed the letter that the Secretary-General had written to him to thank him for the "peacemongering" editorial:

> Let me send you a line thanking you for finally saying a thing that should have been said long before. I refer to the first paragraphs of your editorial, 29 November, and warning against overburdening the United Nations and my office. . . . It is a matter of course that a continued use of the office of the Secretary-General in that way sooner or later leads to a point where he must break his neck, politically.[47]

THE WATERSHED PERIOD, 1959-1964

Two events that occurred in 1959—the accession of Fidel Castro to power in Cuba and the visit to the United States by Nikita Khrushchev—

mark the beginnings of a noticeable divergence between Ascoli and some of his friends and readers over the issue of Communism. In the first case the divergence was not immediately apparent. Just after Castro had come to power, *The Reporter* expressed its confidence "that he will prove to be a vast improvement over his predecessor in the leadership of the island republic."[48] Ascoli soon changed his mind about Castro, however, under circumstances that suggest why he may have felt like an outsider among some members of America's reputed foreign-policy establishment:

> What changed my mind about Castro was a talk he gave here in New York at the Council on Foreign Relations [in the spring of 1959]. He talked as if he had been the leader of Latin America. He talked in the most *arrogant* way to the audience of the Council. . . . They are decent people, professors and bankers and so on, and to my astonishment after we had been told there what he thought of our country and so on they all got up and gave him a standing ovation. And as far as I know only the self-made American born in Ferrara—myself—stood with his ass on the chair. . . . Then we were against him and were even in favor of Johnson's action in the Dominican Republic.

A year later *The Reporter* carried Theodore Draper's critical study of "The Runaway Revolution" in Cuba, and a few months afterward Ascoli wrote, "Cuba is being taken over, not by a flowery imitation of Communism but by the real thing."[49]

Nikita Khrushchev's visit to the United States in 1959 led Ascoli to write the longest editorial that ever appeared in *The Reporter*—it was five pages long—entitled "Now That We've Seen Him—." Although he felt that Khrushchev had run "circles around the American people, and it is to be feared, the high officials of the administration," he concluded that it had been "good to have had him here," for "Khrushchev has reminded those of us who are not satisfied with talk for the sake of talking how excruciating and risky is this business of coexisting with Communism."[50] In the next issue, three letters of reaction were published. Senator Stuart Symington wrote that he considered it a "great editorial"; Paul Nitze called it "one of Max Ascoli's best and most important." But a woman in New York wrote: "I was under the impression that *The Reporter* was a liberal journal. This impression has faded with each issue, and now Max Ascoli's editorial has wiped it out entirely."[51] The eight letters in the issue of November 12, 1959, were divided equally for and against it. The dissenters labeled it "smug" or "intemperate"; one of the four who responded affirmatively called it the "best editorial Ascoli has ever written."

The division between Ascoli and some of his readers over the Khrushchev visit was deepened by a major crisis of the Kennedy Administration—the crisis over Berlin in the summer of 1961. That crisis had been building up for more than two years. On November 10, 1958, Khrushchev had announced his intention of terminating Allied rights in West Berlin by signing a peace treaty with East Germany. Shortly afterward Ascoli wrote a strong editorial, "No Retreat from Berlin" (December 11, 1958), and he continued to urge Allied firmness. In the issue of July 20, 1961, Ascoli wrote: "It is coming. The test, the confrontation between the West and the Communist empire, between our peace and their peace."[52] In the previous issue Eric Sevareid had expressed a similar sentiment: "The showdown with the Communist world conspiracy is on. We have entered the final stage of the long struggle to determine if we can hold our world position short of a great war."[53]

Ascoli normally made a trip to Europe in the summer—*The Reporter* dropped two nonconsecutive issues each summer from its publishing schedule—and he was in Europe when the Berlin Wall was built in August. When a German acquaintance asked him why American journalists and politicians were saying that America had no concern for refugees from East Germany, Ascoli wrote, "seldom have I been so miserable," for he recalled that he himself had been a refugee from tyranny.[54] He felt that the American government should have taken some action—indulging in the traditional privilege of journalists who criticize the government, he did not spell out precisely what it should have been—and in a later editorial he wrote bitterly that "presumably one of the reasons why the administration did not lift a finger against East Germany when the wall was erected may have been the dread that any action on our part against that wretched government might have risked succeeding."[55] This accusation reflects not only the feelings expressed by some West Berliners, [56] but also Ascoli's fear that the liberal policy of "containment" was in practice one of guaranteeing not to interfere with what was happening inside the Communist empire and tacitly to accept its permanency. He feared that Americans, being "a highly traditionalist breed," would see it as only natural that the "cold war"—a metaphor Ascoli always despised—should be followed by some sort of peace that could be reached through negotiations. He did not believe that such a simple resolution of American differences with the Soviet Union was very likely. Toughness was needed. He felt that

> if we act quickly and firmly when the Communists do something like the walling of Berlin, then they will not just stop; they will fall back. The evil which is in them, and which they have wrought upon so many human beings, works in such a way that they must constantly be on the move—forward or backward.[57]

These editorials on the Berlin Wall aroused such strong reactions from Ascoli's personal friends and from readers that he felt compelled to write the first of the "Somewhat Personal" statements quoted in Chapter 2. Of the six letters printed in response to this statement, four supported him, and one—from a Democrat—complained that he had not been critical enough of the New Frontier; but the sixth—from a Texan—said: "Reluctantly I write to you to say that I feel that your magazine no longer expresses and espouses the liberal cause upon which it was founded and has thrived for so many years."[58]

It is a curious fact that this period when *The Reporter*'s anticommunism was becoming increasingly conspicuous was a period of growth in its circulation second only to its initial spurt in circulation in the early 1950s when it was conspicuously anti-McCarthy. Apparently there was a receptive audience among a number of those who considered themselves liberals for *The Reporter*'s combination of liberalism at home and outspoken opposition to Communism abroad. A reader in Connecticut wrote:

> You may be interested to know that several *Reporter* readers with whom I've spoken had identical reactions about the magazine's development. They believe, as I do, that during the past two years or so *The Reporter* has evolved from an ivory-tower preacher's platform to the nation's most effective expositor of the issues of the day.[59]

Similar sentiments were expressed a month later by readers in Berea, Ohio, and Fayetteville, Arkansas.[60]

Other readers strongly objected to *The Reporter*'s views on Communism. John Bennett, dean of the Union Theological Seminary, wrote to *The Reporter* in 1963:

> I hope to live long enough to read that you have come to admit that the changes that have taken place in the Communist countries have some significance for me. . . . The fact of revisionist Communism and the many gradations in the Communist world from Poland to China make constructive coexistence with some Communist countries possible and alter considerably what is at stake in the cold war. It is the complete rejection of this idea in your editorials that has troubled me for some time.[61]

This view of Communism—which represents generally that of many of Ascoli's critics—judges it largely in political terms. But Ascoli, who had long spoken of the need to coexist with Communist countries and was certainly aware of political distinctions among them,[62] judged Communism largely in ethical terms:

> No Communist country has changed its political structure or mellowed its tyranny; witness Yugoslavia. None has recognized man's spiritual life and his separate allegiance both to Caesar and to God, evidence being Poland, one of the countries most patronized by de Gaulle and by us.[63]

Ascoli's continued insistence that "what the people call the cold war is *not* over; neither is Communist enmity toward us" irritated readers who generally respected *The Reporter*'s position on other matters.[64] They commended some of the magazine's critical articles on the Soviet Union; but there were some sacred cows they felt the magazine should leave alone. George Bailey's report on the trial of dissident Soviet writers André Sinyavsky and Yuli Daniel (February 24, 1966) drew an enthusiastic letter of approval from Philip E. Mosely, director of the European Institute of Columbia University.[65] But when the same issue that contained Mosely's letter also contained Bailey's critical article, "Cultural Exchange as the Soviets Use It," Mosely, who had devoted hundreds of hours to the exchange program, wrote a vigorous letter of rebuttal. He twice put Bailey's name in quotation marks, as if to imply that *The Reporter*'s East European correspondent, whom he had so recently praised, had now become a mythical personage. Noting that the article had aroused opposite reactions from a number of qualified readers, Ascoli commented editorially that he was inclined to agree with those who found the article disturbing. "Yet assuming that there can be a two-way improvement in cultural relations with the Soviet Union, an effort of this kind had to be made. It should not be the last."[66]

Ascoli's disagreement with some of his readers over Communism had a philosophical and ethical basis; his disagreement with them over Allied unity had its basis in personalities. In the abstract most liberals agreed with Ascoli on the importance of Allied unity. But when it came down to specifics, there was a real difference, because many liberals were quite unsympathetic with two of the chief West European spokesmen of the late 1950s and early 1960s—Konrad Adenauer and Charles de Gaulle. Although he had never met either of them—to Ascoli's great disappointment, several arrangements for a meeting with de Gaulle had fallen through—he liked and respected them both. In one editorial he described them as "two patriarchs of superb intelligence and will power, who happen to be well above their people in terms of moral stature and of purposeful energy. They compensate for the past and present deficiencies of their countrymen."[67]

When de Gaulle was voted into office in 1958, Ascoli wrote: "Strangely enough, the fate of western democracy, at least on the European continent, maybe even the survival of the western coalition, largely depends on the success of this extraordinary Frenchman, Charles de Gaulle."[68] The following year he noted—perhaps more prophetically than he intended—"Some of his designs, if carried through, may shock us."[69]

Ascoli fully recognized de Gaulle's phenomenal ability to rub people the wrong way, but he also believed that he was "a man who throughout his long momentous life has never stopped growing. He is now a soldier-statesman who has come to recognize the obsolescence of militarism, of war, and of colonialism."[70] Ascoli and de Gaulle shared the same conception of the West; de Gaulle wrote of his final conversation with President Roosevelt:

> "It is the West," I told President Roosevelt, "that must be restored. If it regains its balance, the rest of the world, whether it wishes or not, will take it as an example. If it declines, barbarism will ulti-

> mately sweep everything away. Western Europe, despite its dissensions and its distress, is essential to the West. Nothing can replace the value, the power, the shining examples of these ancient peoples."[71]

Recognizing that de Gaulle needed "our understanding and patience," Ascoli wrote in the last year of the Eisenhower Administration: "We seem ever ready now to negotiate with our potential enemies. Could we not give some evidence of patience, of understanding, in working with our friends?"[72] In the last year of the Kennedy Administration, he was still making the same plea:

> In dealing with the nations of our community, we seem to be singularly intolerant of opposition from those who stand up to us. . . .
>
> This is eminently the case with General de Gaulle. He is, to be sure, an uncomfortable, stubborn, self-centered man, sometimes incredibly wrong. Yet there is no greater man among statesmen alive today, or one who cares more deeply for the ideals we believe in.[73]

In April 1963, Ascoli attacked the Kennedy Administration's use of the press to conduct an "ever mounting campaign against de Gaulle, in which all means have been used, including deliberate misinformation."[74] A reader in Arizona contributed this thoughtful response:

> It is certainly true that most of our organs of information, including quite a few that do not normally support members of Mr. Kennedy's political party, have tended over the past several months to portray President de Gaulle as a self-centered, obstructionist old man whose antics were nonetheless being tolerated with amazing patience by the wise young fellows in the White House. Having in large measure shared this view myself, I would like to take this opportunity to congratulate you for helping open my eyes to at least some of the quite reasonable concerns that must be on the minds of many thoughtful Europeans. I would like to cite in addition to your editorials the excellent articles by Henry A. Kissinger in your March 28 issue and by Eugene V.

Rostow in the issue of April 25. The problems are indeed complicated, and I am not at all ready to say that de Gaulle has all the right answers by any means, but you have at least made it clear to me that we must pay attention to what he says.[75]

The Reporter's impatience with the clichés employed by both liberals and conservatives concerning foreign policy in the last years of the Eisenhower Administration and during the Kennedy Administration is perhaps conveyed most vividly—and certainly most amusingly—in the pages of Meg Greenfield's satiric fantasy, "The Seattle Settlement" (May 24, 1962). In it she describes a U. S.—Soviet agreement to neutralize Seattle with an American mayor, Soviet police chief, and uncommitted city manager, and then describes liberal and conservative reactions. It is a brilliant parody of some of the foreign-policy debates of the period; the exaggerated arguments she puts into the mouths of spokesmen for each side are hilarious, but they have at the same time the uncomfortable ring of authenticity.

Ascoli's criticism of liberal clichés became more pronounced after Lyndon Johnson became President. A few months after Kennedy's assassination Ascoli took to task three old friends—he called them "all good men"—J. William Fulbright, McGeorge Bundy, and Adlai Stevenson, who seemed to him to be attempting, perhaps unconsciously, "to codify the Kennedy legacy," which he described as "a legacy of all-around goodwill, many laudable intentions and few achievements." He felt they had an unjustified optimism about the changes taking place in the Communist world: "Is there a single Communist-ruled society that offers any indication or even any slight hope that it will break open?"[76]

VIETNAM

The culmination of *The Reporter*'s controversial stand on foreign policy was its support of the Vietnam War—support that estranged many of Ascoli's liberal friends and a number of his readers and that played a part in precipitating Ascoli's abrupt announcement in April 1968 that the magazine would close two months later. *The Reporter*'s position on Vietnam is conspicuously marked with ironies, for the Vietnam war progressively demonstrated the disastrous consequences of several weaknesses in American foreign policy that Ascoli had persistently exposed and tried to root out over the years: a dangerous American ignorance about the rest of the world; a temptation to become too intimately involved with dubious client governments or leaders; and an inclination to pursue policies that might isolate America from its Western allies.

Ascoli had long been aware of the way political ignorance had hampered American efforts to help other peoples. He was particularly aware of American shortcomings in dealing with his native country after World War II:

> It was not lack of sympathy for the Italian people that marred the Allied policy in Italy nor was it any particular sinister intentions. Rather, there, as elsewhere, there was no clear thought-out policy, based on a detailed knowledge on the part of the Allies, of the problems they had to face and of the responsibilities they had to assume.[77]

One of his articles on postwar reconstruction in Italy contains two prophetic sentences: "It is a difficult art, that of stimulating and at the same time respecting the freedom of liberated countries. To a very large extent, peace depends on the acquisition of this skill by the largest possible number of responsible men."[78]

The Reporter's first article on Vietnam appeared when the magazine was less than a year old; from 1950 on it carried an average of between two and three articles yearly until 1965, when its coverage, like American military involvement, greatly escalated. Many of these first forty articles, like most of the subsequent ones, were written by *The Reporter*'s correspondent in Southeast Asia, Denis Warner. Occasional contributors included Joseph Buttinger, Peggy Durdin, Bernard Fall, Stanley Karnow, Wolf Ladejinsky, S. L. A. Marshall, Robert Shaplen, and Theodore H. White. The first two articles published by *The Reporter* on Vietnam bear titles that could summarize the debate of the next two decades: "Viet Nam—Roadblock to Communism" (February 28, 1950) and "A New Disaster in Asia?" (April 11, 1950). The first article, by Samuel G. Welles, defended the American policy of recognizing the Bao Dai regime in South Vietnam and of providing military and economic assistance. Welles recognized the problem of leadership in that country but argued that Bao's regime was essentially nationalistic rather than French, and that American assistance had a good chance of enabling South Vietnam to avoid the extremes both of Communism and of colonialism. The second article, by Harold Isaacs, presented the argument against American support of the French and of the Bao Dai regime. Isaacs argued that "there is not a single American official of policy-making rank in the State Department today who is in a position to view Asian problems from any kind of mature background, knowledge, and experience with Asian affairs and Asian peoples." He recalled that no foreign army in the twentieth century had been able to win and hold any significant territory in Asia, and he argued that by associating itself with French imperialism America was isolating itself from "those Asian nationalists who most want to be America's friends."

The editorial in the issue that contained the Isaacs article disagreed with him and supported the United States' policy in Vietnam. It attacked a liberal cliché about nationalism:

> We should avoid looking at the nationalism of other nations with a sort of curious but widespread racial bias: the assumption that nationalism is a vice for white nations and a virtue for the colored ones. When we call De Gaulle a nationalist we imply, of course, that he is no good; but if Ho Chi Minh claims to be a nationalist—Oh, well!

In Indo-China, *The Reporter* admitted, the United States was taking a gamble, just as it had earlier when it had intervened in Iran, Greece, Turkey, and Indonesia, where conflicts developed that "could have sparked a new international conflagration" and where "our government could not help intervening and had no choice as to which side it was going to back." Indo-China represented "probably one of the most reckless gambles we have ever had to take. The available reports on Indo-China are, to say the least, contradictory." Yet there really seemed to be no alternative:

> In Indo-China we have had to make a loathsome choice. His former Imperial Majesty, Bao Dai, is by no means a savory character. But if our representatives make Indo-Chinese autonomy the condition for our aid, both the French and the Indo-Chinese people will be the gainers.[79]

All of *The Reporter*'s articles on Indo-China during the 1950s—even the most optimistic, Joseph Buttinger's "An Eyewitness Report on Vietnam" (January 27, 1955)—displayed a vivid awareness of the dilemma facing United States policy in Southeast Asia. At the beginning of the decade Robert Shaplen wrote: "As part of our necessarily global policy, military aid to France in Indo-China is essential." But the question remained: "How much social and economic good can we still do in ratio to the 'harm' our military-aid program causes in terms of Vietnamese lives and homes destroyed by U.S. weapons?"[80] At the end of the decade Wolf Ladejinsky, who had been an adviser to the South Vietnamese government on agricultural policy, wrote: "Vietnam's dilemma is how to keep its security force unimpaired while making decisive economic strides. This is also America's dilemma. . . . And the dilemma cannot be resolved by playing down one priority in favor of another."[81] In 1958 *The Reporter* carried a report on the failure of United States aid policies in Laos, written by a former employee of the International Cooperation Administration, Haynes Miller, who had worked in that country:

> Instead of creating a situation of military strength as a "bulwark against Communism," our policy has made Laos into what a recent report of the General Accounting Office has called "a financial dependency" of the United States. It would be bad enough if the money had been merely wasted. The record to date, however, shows that our Laotian policy has demoralized our friends and strengthened our enemies.[82]

After 1965, like the rest of the news media, *The Reporter* gave increasing attention to Vietnam, and in 1966 it published *Vietnam: Why*, a fifty-six page collection of articles and editorials from its 1965 and 1966 issues. The book was originally prepared by Shirley Katzander as a promotion item to be distributed to advertisers. When someone suggested selling copies for a dollar apiece, the circulation manager gloomily predicted they would be lucky to sell fifty. But, promoted solely through full-page advertisements in *The Reporter*, it sold 20,000 copies within a few months. Orders came from every state in the Union, from a Swedish radio commentator, an Oxford don, and a Greek businessman in Athens, among others. The White House ordered five; the Navy ordered six hundred; the Hawaii office of the Agency for International Development asked for fifty. The

Central Intelligence Agency asked for sixteen—by special delivery—and then ordered twenty more.[83]

One of *The Reporter*'s persistent warnings on foreign policy over the years concerned the hazards of too-intimate relations with governments that were recipients of United States assistance. Ascoli had opposed any alliance with Franco; he persistently argued against too close an association with Chiang Kai-shek; and he warned against treating Tito either "with haughty righteousness or with mushy intimacy."[84] Uneasy about Bao Dai, *The Reporter* was pleased when Ngo Dinh Diem became Premier of South Vietnam. Referring to Diem as "a quiet little man we once knew in New York," an unsigned item in "The Reporter's Notes" commented: "Only time can tell . . . but, considering that since the war we have been backing one Darlan after another, we certainly felt good when we learned that even in as explosive a country as Vietnam the experiment of backing an honest man is being tried."[85]

After Diem's assassination, another unsigned item in "The Reporter's Notes" commented:

> Many of Diem's difficulties, no doubt, stemmed from destructive inner conflicts. But it must also be assumed that the United States, which could legimately claim to share the credit for Diem's initial successes, must share some of the credit for his ruin and downfall. Official disclaimers notwithstanding, the coup in Saigon and all that immediately flows from it are a direct responsibility of the United States government.

The note concluded: "Above all, let it be clear that we cannot stay in Vietnam forever with the power of life and death over the country and its leaders."[86]

Ascoli's most revealing argument in support of American involvement in Southeast Asia—despite its many admitted failures—was written in 1958, before his defense of United States policy became inextricably intertwined with his fight against the Vietnam protesters at home. In an editorial accompanying the Haynes Miller article on Laos, Ascoli wrote:

> Communism commits us to the assistance of nominal democracies in nominal nations all over the world. . . .
>
> The case of Laos is no isolated one, and the conclusion is invari-

> ably the same everywhere: Communism is the evil beneficiary of all the mistakes we make. Yet it must be added most emphatically that if we have produced, or sponsored, or failed to prevent messy situations in too many of the countries we have been trying to assist, *even this messiness is better than withdrawal into Fortress America.*[87]

Isolationism, then, was the supreme folly for America. Moreover, as he warned in the same editorial, America needed "the closest possible cooperation with our allies" because we had learned through our mistakes in Asia "that we are not powerful or experienced or wise enough to act alone."

It hardly needs to be pointed out, however, that Ascoli himself rejected the advice on Vietnam given by his hero Charles de Gaulle, or that in trying to avoid military isolation America found itself at odds with its allies and headed toward a kind of psychological isolation. *The Reporter*'s stand on Vietnam, then, is marked by contradictions, paradoxes, and irony. This is hardly surprising, since American involvement in Vietnam was marked by the same qualities:

> One ironic aspect of the war in Vietnam is that while we possess an idealistic philosophy, our failures have been due to an excessive reliance on material factors. The Communists, by contrast, holding to a materialistic interpretation, owe many of their successes to their ability to supply an answer to the question of the nature and foundation of political authority.[88]

A perceptive and prophetic description of America's most fundamental problem in Vietnam was once given by Ascoli:

> Revolutions cannot be made by foreign governments, no matter how friendly and well-intentioned. . . . Revolutions must be made by the people themselves. . . . But there is no revolution conceivable when a people has lost the sense of its own institutions and the capacity to recognize itself through its own leaders.

This observation was not made about Vietnam. It was written in 1944 about Allied efforts to reconstitute a government in a war-torn Italy.[89]

It finds an echo in a remark of "a cynical and frustrated Vietnamese friend" quoted by Denis Warner in 1966: "The Americans can teach us a lot, but they cannot teach us how to love our country."[90]

The Reporter, of course, was far from alone in supporting United States involvement in Vietnam, particularly prior to 1965. In the early months of the Kennedy Administration the *New York Times*, soon to become a critic of the war, carried an editorial entitled "Subversive Aggression" (April 14, 1961) that boldly if vaguely urged that something be done about Communist aggression in South Vietnam:

> The free world must unceasingly protest against and oppose Communist subversive aggression, as practiced most acutely today in Southeast Asia. To accept it as a matter of course is to hand the Communists half a victory without a fight.

And after the *Times* had turned against the war, the *Washington Post*, one of the most outspoken liberal newspapers in the country, continued to support it—not always with much enthusiasm—for several years.

In continuing to support United States involvement in Vietnam after other liberal publications had turned against it, *The Reporter* illuminates a fundamental dilemma of a liberal foreign policy. "The danger," Harlan Cleveland had warned in a 1953 editorial on Southeast Asia, "is that we shall not stay the course, that we shall get discouraged when our policy fails in every instance to produce the advertised results."[91] Fifteen years later *The Reporter* still seemed to be heeding that warning: if American aid to Vietnam was right in principle, then the only proper course was to persist until it produced the desired results. But was it right in principle? That was the liberal dilemma, and here the liberal desire to have things both ways was hardly helpful. Ascoli, who was not wholly free from this liberal desire, could in the same editorial in 1955 argue against the United States adopting "a sort of global Monroe Doctrine" and at the same time urge:

> We can also show by our deeds—our own and those of the United Nations—how assistance can be given to hard-pressed peoples struggling against old and new colonialisms.
>
> In South Vietnam, for instance—and right now.[92]

The Reporter's editorials on Vietnam reflected two elements that have traditionally played a role in the formulation of American foreign policy—national self-interest and loyalty to universal ideals. Horton's occasional editorials, such as "The President's Decision" (February 24, 1965) and "The Great Debate" (March 11, 1965), tended to emphasize arguments based on self-interest. Ascoli's, by contrast, tended to emphasize the arguments based on idealism, although he was also aware of the component of self-interest. "Because of the American fighting soldiers," he wrote in 1966, "that poor battered Asian country has become the pivot in our—and not just our—contest with Communism."[93] But he continually felt that the central issue was a moral one, arguing in one editorial that "lack of concern on our part for the people of Asia . . . is immoral, for the notion is immoral that the men and women of Asia are a separate breed, a sort of half-caste world, for whom a system of government and values that Western civilization abhors is good enough."[94]

In the last two or three years of *The Reporter*'s existence Ascoli may have felt trapped, just as America seemed to be trapped in Vietnam. Sophisticated enough to be fully aware of Communist polycentrism, he was at the same time moralist enough to believe that Communism remained basically evil—particularly in its denial of civil liberties. And he continued to disagree, as he had in 1951, with "the traditional journal-of-opinion liberalism" which he described as consisting of "the firm belief that no American policy could be right."[95] He felt that opposition to President Johnson's Vietnam policy represented both a return to isolationism and a "revolt against freedom." And as it became more and more fashionable among intellectuals to attack Johnson, Ascoli—who always hated intellectual fads—grew only the more determined to support him.

"We found ourselves in South Vietnam because of the accidents of diplomacy," Ascoli wrote early in 1966. "Now we are there because we must."[96] He saw no way the United States could evade the obligations it had incurred:

> We have fallen into too many miscalculations and blunders in endeavoring to assist the people of that tragic land. But the way Communism has robotized men and women in the Vietminh and the Vietcong is so horrid that as long as there are people in South Vietnam who, no matter how distraught by suffering and intrigues, still don't want to give up, we cannot possibly tell them that, much to our regret, we are leaving.[97]

"Negotiations" seemed to him to provide no easy way out, for the example of Laos "has proved to be too disastrous to make a second and larger round even thinkable."[98] The problem was larger than Vietnam; it was our whole relationship with the "two Communist empires" which "still gallantly co-operate in doing harm to us." Despite talk about the end of the cold war, a "bloody clash was bound to occur, for a reconsideration of our relations with Communism had become imperative. It happened in Vietnam." And Ascoli recognized that the stakes at home and abroad were high:

> There is no denying that the danger we are running is exceptionally great. In this era of war games as an avoidance of total conflagration, our country has fought in Vietnam a limited war conceived as a substitute for a preventive one. It may turn out to be the closest equivalent to a total war. If our institutional structure, if all the links between races, group interest, and national tradition snap under the stress of unchecked fury, then we are through.[99]

"The fundamental trouble," Ascoli wrote in 1967, "does not come from Communist aggressiveness or Communist wiles but from the complacency of so many Americans and so many allies. Such complacency, sometimes mixed with muddleheadedness or with fear, can make a powerful country into a docile satellite." He believed that the "great confusion and bewilderment" in the country were caused only incidentally by Vietnam. He traced them back to an unhealthy combination in American thinking of "lofty universal principles" and "that tricky, instance-by-instance practicality called pragmatism." He could find no other word for the "current agitation, adolescent or senile" in the country, but "nihilism," and he feared a trend toward "the absolute power of a mob first, and then of a tyrant." [100] Those most to blame were those whom he had analyzed thirty years earlier in his book *Intelligence in Politics*—the intellectuals. In one grim glimpse of a possible future, he suspects that they may be among the tyrants:

> Sooner or later there will be a worldwide sameness; a jet set dedicated to the drinking of Bloody Marys while the ruling ideas are provided by a demi-monde of demi-intellectuals—in other words the same jet set. At the very bottom of society there will be the kept poor, beneficiaries not of charity but of a guaranteed income.[101]

Thus in the end, as Ascoli had always maintained, there was no line between foreign affairs and domestic affairs. In both there was the political problem of making—in Ascoli's phrase—"freedom operational." And behind this lay an ethical question: How do people make the right use of their freedom, so that they do not use freedom to destroy freedom? In the climate that pervaded the country after the middle of the 1960s, Ascoli began to doubt whether he could articulate an answer that would be heeded. By the spring of 1968, he recalled three years later:

> I felt I was stuck to a completely quixotic position. I have found a situation—and unfortunately I have been too good a prophet—in which I did not see the end. Not that I was a warmonger, although I never denied—different from many of my friends—the importance of the war in Vietnam because the more it was attacked, the more it exposed the flank of American policy and of American existence even.

He realized that "something was radically changed in America, and the injection of liberalism in my sense affecting a certain number of intellectuals or semi-intellectuals was not enough." He was not only estranged from many of his friends; he also felt—wrongly, as it turned out—that he had lost hold on his readership; he knew there was rebellion on the staff. "I couldn't run a magazine under those conditions." And so, on April 9, 1968, a week and a half after President Johnson's withdrawal speech, Ascoli formally announced that *The Reporter* would cease publication with the issue of June 13, 1968.

NOTES

1. *The Reporter*, June 6, 1954, inside front cover.
2. Walter Millis, "A Liberal Military-Defense Policy," in *The Liberal Papers*, ed. James Roosevelt (Garden City, N. Y.: Doubleday, 1962), pp. 97, 119.
3. Quincy Wright, "Policies for Strengthening the United Nations," in *The Liberal Papers*, p. 316.
4. Alexis de Tocqueville, *Democracy in America*, ed. Phillips Bradley (New York: Vintage, 1958) vol. 1, p. 243.
5. It was Thomas Jefferson, patron saint of liberals and probably our most cosmopolitan President, who warned Americans in his first inaugural address against

"entangling alliances." A similar warning—though not the phrase itself—looms large in Washington's farewell address.

6. Quoted in Robert E. Osgood, *Ideals and Self-Interest in America's Foreign Relations* (Chicago: University of Chicago Press, 1953), p. 300.

7. Osgood, pp. 370-71.

8. Francis E. Rourke, "The Domestic Scene," in *America and the World*, ed. Robert E. Osgood (Baltimore and London: Johns Hopkins Press, 1970), p. 178.

9. See e.g., his critical review of *Witness*, "Lives and Deaths of Whittaker Chambers," *The Reporter*, July 8, 1952, reprinted in *Our Times*, pp. 162-71.

10. See André Sinyavsky [Abram Tertz], "On Socialist Realism," in *The Trial Begins and On Socialist Realism* (New York: Random House, Vintage, 1960), esp. p. 162, and Jacques Ellul, *The Technological Society* (New York: Random House, Vintage, 1964).

11. Ascoli, "Station Identification," *The Reporter*, September 26, 1950, p. 1.

12. Ascoli, "War Aims and America's Aims," *Social Research* 8 (September 1941), pp. 270, 271.

13. Ascoli, "Must We Have Compulsory Arbitration?" *The Reporter*, November 12, 1959, p. 15.

14. Ascoli, "If We Must Have Spain. . . ," *The Reporter*, June 20, 1950, pp. 4-5.

15. Ascoli, "Fascism in Our Neighborhood," *The Reporter*, February 28, 1950, p. 5.

16. Ascoli, "Coalition Government in Italy," *Free World*, June 1944, p. 516.

17. Ascoli, "Can We 'Co-Exist' with Malenkov?" *The Reporter*, April 14, 1953, p. 9.

18. Ibid.

19. Ascoli, "Breaking the Russian Spell," *The Reporter*, October 7, 1954, pp. 12-13.

20. Ascoli, "From Utopia to Reality," *The Reporter*, November 22, 1955, p. 12.

21. Ascoli, "Their Madness and Ours," *The Reporter*, June 23, 1960, p. 15.

22. Ascoli, "Escalation from the Bay of Pigs," *The Reporter*, November 8, 1962, p. 25.

23. Ascoli, "Can We 'Co-Exist' with Malenkov?" *The Reporter*, April 14, 1953, p. 9.

24. Ascoli, "America, the Vatican, and Israel," *The Reporter*, January 22, 1952, p. 5. Emphasis supplied. Ascoli was not always this careful to distinguish between America as a pattern for mankind and America as representing the potential for mankind's development.

25. Ascoli, "The Two New Faces of Janus," *The Reporter*, July 9, 1959, p. 8.

26. Ascoli, "Matrix of Commonwealths," *The Reporter*, July 12, 1956. Cf. Ascoli, "Three Countries—and Us," *The Reporter*, June 23, 1953, p. 7.

27. Ascoli, "Civilization *Is* the West," *The Reporter*, December 27, 1956, p. 6. Emphasis in the original. Cf. "Can We Make Peace with Our Allies?" *The Reporter*, June 7, 1962, p. 14, and "Back from Europe," *The Reporter*, September 13, 1962, p. 21.

28. Ascoli, "The Future of the U.N.," *The Reporter*, October 26, 1961, p. 23.

29. Ascoli, "To Stop This Madness," *The Reporter*, January 4, 1962, p. 13. Emphasis in the original.

30. J. William Fulbright, "For a Concert of Free Nations," *Foreign Affairs* 40 (October 1961), 1-18.

31. *Nation*, December 17, 1955, p. 525; January 21, 1956, p. 41; August 18, 1956, p. 129; *New Republic*, April 15, 1957, p. 3; October 6, 1958, pp. 3-4.

32. Ascoli, "It Happens in California," *The Reporter*, February 21, 1957, p. 12.

33. Ascoli, "Our Man," *The Reporter*, February 20, 1958, p. 2.

34. Ascoli, "The Mandate: A Coalition Cabinet," *The Reporter*, November 15, 1956, p. 10

35. Ascoli, "Peace–the Forbidden Issue," *The Reporter*, October 18, 1956, p. 10.

36. Ascoli, "The Ordeal of Mr. Dulles," *The Reporter*, March 5, 1959, p. 10.

37. Ascoli, "Stalin Makes Up His Mind," *The Reporter*, May 13, 1952, p. 6.

38. Ascoli, "Off to a Good Start," *The Reporter*, July 14, 1955, p. 8.

39. Ascoli, "The Ordeal of Mr. Dulles," p. 10. In 1971 Ascoli conceded privately, "I have probably been a little bit too hard on John Foster Dulles."

40. Ascoli, "Foreign Policy after Cuba," *The Reporter*, May 25, 1961, p. 19; "China's New Leap Forward," *The Reporter*, September 23, 1965, p. 24; "On Hawks and Doves," *The Reporter*, March 24, 1966, p. 24. For a concurring view of Rusk's personal qualities by a liberal who worked for him briefly but did not agree with him on Vietnam, see Charles Frankel, *High on Foggy Bottom* (New York: Harper and Row, 1969), pp. 226-27.

41. Ascoli, "A Plea for Recognition of the U.S.," *The Reporter*, January 3, 1950, pp. 2-3.

42. Ascoli, "For a Formosa Settlement" and "As of Now," *The Reporter*, October 16, 1958, pp. 10-17.

43. Ascoli, "The Price of Peacemongering," *The Reporter*, November 29, 1956, p. 10.

44. *The Reporter*, February 7, 1957, p. 8.

45. Ascoli, "In Defense of the U.N.," *The Reporter*, February 7, 1957, p. 22. This article was part of a friendly debate with James Warburg over United States policy toward the Middle East (February 7 and February 21, 1957).

46. Ascoli, "The Future of the U.N.," *The Reporter*, October 26, 1961, p. 21.

47. The letter is quoted at length in "The Future of the U.N.," Ascoli's three-page editorial tribute to Hammarskjöld.

48. "After the Haircut," *The Reporter*, January 22, 1959, p. 2.

49. Theodore Draper, "The Runaway Revolution," *The Reporter*, May 12, 1960, pp. 14-20; Ascoli, "Latin America Joins the World," *The Reporter*, September 15, 1960, p. 16.

50. Ascoli, "Now That We've Seen Him–," *The Reporter*, October 15, 1959, pp. 21-22.

51. *The Reporter*, October 29, 1959, p. 8.

52. Ascoli, "The Long Moment of Truth," *The Reporter*, July 20, 1961, p. 18.

53. Eric Sevareid, "The Facts of Life," *The Reporter*, July 6, 1961, p. 13.

54. Ascoli, "The Wall," *The Reporter*, September 14, 1961, p. 22.

55. Ascoli, "This 'Red-or-Dead' Nonsense," *The Reporter*, October 12, 1961, p. 27.

56. See Mayor Willy Brandt's letter to President Kennedy, calling for "not merely words but political action" against the closing of the border between East and West Berlin. (*New York Times*, August 17, 1961, p. 1.)

57. Ascoli, "This 'Red-or-Dead' Nonsense," pp. 27-28.

58. *The Reporter*, December 21, 1961, p. 6.

59. *The Reporter*, September 13, 1962, p. 10.

60. *The Reporter*, October 11, 1962, p. 6.

61. *The Reporter*, June 20, 1963, p. 8.

62. See, e.g., Ascoli, "The Leaders and Their Mandates," *The Reporter*, November 5, 1964, p. 16.

63. Ascoli, "Maître de Gaulle," *The Reporter*, June 30, 1966, p. 10. Because he believed Communism to be inherently antithetical to civil liberties, Ascoli was always alert to evidence—and it kept turning up—that would corroborate his belief. Some of the problems the Western reporter has in conveying the amount of repression practiced in a country like Poland are discussed in David Halberstam's thoughtful article, "Love, Life, and Selling Out in Poland," *Harper's*, July 1967, pp. 78-89.

64. Ascoli, "The Prodigal Nation," *The Reporter*, September 8, 1966, p. 23.

65. *The Reporter*, April 7, 1966, p. 8.

66. *The Reporter*, May 19, 1966, p. 11.

67. Ascoli, "De Gaulle the Indispensable," *The Reporter*, February 18, 1960, p. 14. George Bailey, *The Reporter*'s correspondent in Germany, knew Adenauer personally. Able to drop into Adenauer's office and converse with him in fluent German, Bailey called him "the most accessible politician I have known" and wrote an eloquent tribute to him after his death. (Bailey, "Konrad Adenauer," *The Reporter*, May 4, 1967, pp. 12-13.)

68. Ascoli, "Once More the Heart of Europe," *The Reporter*, June 26, 1958, p. 8.

69. Ascoli, "Let's Pretend," *The Reporter*, December 10, 1959, p. 15.

70. Ascoli, "Kennedy and De Gaulle," *The Reporter*, December 10, 1959, p. 15.

71. Quoted in David Calleo, *The Atlantic Fantasy: The U.S., NATO, and Europe* (Baltimore and London: Johns Hopkins Press, 1970), p. 58.

72. Ascoli, "De Gaulle the Indispensable," *The Reporter*, February 18, 1960, p. 14.

73. Ascoli, "As If We Were at Peace," *The Reporter*, January 31, 1963, p. 23.

74. Ascoli, "A Few Anguished Questions," *The Reporter*, April 25, 1963, p. 22.

75. *The Reporter*, May 23, 1963, p. 6.

76. Ascoli, "Public Diplomacy," *The Reporter*, April 9, 1964, p. 10.

77. Ascoli, "After the Fascist Hoax," *Free World*, February 1946, p. 24. Cf. an earlier and more scholarly article by Ascoli, "The Lesson of Italy," *Social Research* 11 (May 1944), 135-51.

78. Ascoli, "Italy, An Experiment in Reconstruction," *Annals of the American Academy* 234 (July 1944), 41.

79. "The Indo-China Gamble," *The Reporter*, April 11, 1950, p. 1.

80. Robert Shaplen, "Indo-China: The Eleventh Hour," *The Reporter*, October 2, 1951, p. 10.

81. Wolf Ladejinsky, "Vietnam: The First Five Years," *The Reporter*, December 24, 1959, p. 23.

82. Haynes Miller, "A Bulwark Built on Sand," *The Reporter*, November 13, 1958, p. 16.

83. Philip M. Dougherty, "Advertising: Vietnam Book Stirs Response," *New York Times*, January 25, 1967, p. 54; *The Reporter*, February 9, 1967, pp. 10, 12; interview with Shirley Katzander, January 1971.

84. Ascoli, "Our Belgrade Gamble," *The Reporter*, February 5, 1952, p. 5.

85. "Little Mr. Diem," *The Reporter*, December 2, 1954, p. 4.

86. "On Playing God," *The Reporter*, November 21, 1963, pp. 14,16.

87. Ascoli, "The 'Sister Republics' of Asia," *The Reporter*, November 13, 1958, p. 10. Emphasis supplied.

88. Henry A. Kissinger, *American Foreign Policy: Three Essays* (New York: Norton, 1969), p. 106. Cf. Douglas Pike, *Viet Cong* (Cambridge, Mass.: MIT Press, 1966).

89. Ascoli, "Coalition Government in Italy," *Free World*, June 1944, p. 515.

90. Denis Warner, "Behind the Battlefront: A Search for Stability," *The Reporter*, February 24, 1966; reprinted in *Vietnam: Why*, p. 39.

91. Cleveland, "The Defense of Asia," *The Reporter*, April 28, 1953, p. 11.

92. Ascoli, "The New Great Debate," *The Reporter*, January 27, 1955, p. 10.

93. Ascoli, "The President Answers," *The Reporter*, July 14, 1966, p. 12.

94. Ascoli, "Separate Worlds," *The Reporter*, July 1, 1965, p. 12.

95. Ascoli, "The Diplomat and the Dinosaur," *The Reporter*, November 13, 1951, p. 11.

96. Ascoli, "Suspense," *The Reporter*, January 13, 1966, p. 20.

97. Ascoli, "The Peace Table," *The Reporter*, December 2, 1965, p. 20.

98. Ascoli, "Negotiations with China," *The Reporter*, January 28, 1965, p. 18.

99. Ascoli, "We and They," *The Reporter*, November 30, 1967, pp. 12-13.

100. Ascoli, "We and They: II," *The Reporter*, December 14, 1967, pp. 12-13.

101. Ascoli, "That Living Anachronism," *The Reporter*, April 6, 1967, p. 10.

Chapter Six

The End of an Experiment

Somehow, I feel that this is the last lecture in the faculty I established nineteen years ago. . . . But if the work of The Reporter *is to be meaningfully concluded, I must close* The Reporter.

Max Ascoli, "Farewell to Our Readers"

The Reporter *has been required reading for me for so long I can no longer remember when I first subscribed. You helped me with my personal growth and made me more effective in the work I do. I feel as though I were losing a patient, kindly, and wise teacher.*

A newspaperman in Akron, Ohio, 1968

The death of an institution is a painful experience. Only those who have lived through the dissolution of an organization that has given them financial and emotional security, and has perhaps embodied their dreams as well, can fully know the heartbreak involved. For several years after *The Reporter* closed, Ascoli and a staff of four—receptionist, librarian, and two secretaries—continued to occupy a portion of the magazine's former quarters at 660 Madison Avenue in New York. Various former staff members would call or drop in from time to time, but there were also former staff members who never called or visited. And among some of the latter there were strong feelings about the demise of *The Reporter*. Ascoli, quite understandably, regarded *The Reporter* as *his* magazine; a number of his staff, just as understandably, regarded it as *their* magazine. And the two conceptions of the magazine were not the same.

Such a difference in views is hardly surprising. Any worthwhile organization serves two purposes: first, it embodies some idea or principle or carries out some function outside itself; and second, it seeks to perpetuate itself. The two purposes are mutually dependent; an organization needs a reason for being, and it needs to survive if it is going to fulfill its function. But they can also become antithetical: the price for survival may seem to involve the sacrifice of the principle that the organization embodies; and if an organization loses its raison d'être, what is the value of its mere survival? In practice, of course, many organizations attain some form of compromise, and in the absence of any real crisis can continue to function acceptably.[1]

But in a time of crisis, financial or otherwise, an organization becomes a battleground, not only between those who have differing conceptions of its basic principles, but also between those who have differing strategies for ensuring its survival. At one extreme are those who would rather sacrifice the organization than sacrifice what they regard as its basic principles; at the other are those who in their desire to perpetuate the organization seem prepared to compromise the principles that have given it meaning. The battleline between these two extreme positions is never drawn with complete clarity, of course, because there are always differing interpretations of the issues and the risks involved. The outsider has difficulty discerning the truth, for the participants almost always have partial views. What honestly appears to one group to be a fight for principles appears to others as egoism, vanity, or stubbornness. What appears to another group as a perfectly reasonable willingness to compromise appears to its opponents as nothing less than a sellout. The judgments of all participants are likely to be affected by such factors as desire for job security, unwillingness to admit mistakes, frustrated ambition, hurt feelings, and simple clashes of personalities. The most tragic aspect of the death of an institution is the division and bitterness generated among those who had thought of themselves as united in a common enterprise.

The question "Why did *The Reporter* fold?" has one obvious answer: its founder, editor, and patron, Max Ascoli, decided to bring it to an end. No former staff member would dispute this answer, but many of them would insist that it only conceals a more basic question, to which the answer is not so obvious: "Was it necessary for *The Reporter* to fold?" Some former staff members were bitter about Ascoli's decision, because they interpreted his loyalties differently than he did. They thought his primary loyalty should have been to the organization and the people who helped him create it; he saw his primary loyalty as being to the idea that he felt the organization embodied. He did not believe that the magazine could have remained true to its essential idea—he described it in 1971 as "that blend of European and American liberalism"—without his guiding presence. He said that over the years he tried to find someone who would be able to take over the magazine and keep it true to its basic principles, and that he was unsuccessful. "Or maybe I am too much of an egocentric," he conceded. Of his staff he said:

> Although I would not have left *The Reporter* to any one of them or to all of them together, I am very grateful for what they have done for me. They may feel resentful toward me because some of their ambitions may have been thwarted. . . . If they hate my guts, that's their privilege. I don't give a hoot in hell. And I don't feel obliged in any way to reciprocate.

Believing that the magazine could have continued only at the risk of sacrificing the idea that gave it meaning, he finally decided that he had to sacrifice the magazine.

Such decisions, of course, are not reached suddenly, nor are they reached in a vacuum. There were objective factors that played a part in shaping his decision. Three factors that were commonly cited in news stories about the magazine's imminent demise were noted by the *New York Times*:

> Financial problems, the desire of Dr. Ascoli, its founder and head, to devote himself to writing, and a declining rate of renewal subscription, possibly reflecting the liberal publication's support of President Johnson and the Vietnam war, were cited as reasons for closing down with the June 13 issue.[2]

Time and *Newsweek* both mentioned the magazine's financial losses and its support of the Vietnam War as factors in its closing. The former noted that advertising pages had dropped from 543 ("moderately money-losing") in 1963 to 401 ("painfully money-losing") in 1967. The latter claimed that most of *The Reporter*'s circulation rise had been "revolving door" readership, consisting of subscribers who tried the magazine for a year or two and then dropped it. *Time* felt that the magazine's demise was the result of its consistency:

> *The Reporter* magazine presented a steadfast face to the world. From the day it started in 1949, its standards of journalism were high, its contributors stayed close to the facts, and it enthusiastically accepted the postwar role of the U.S. as a world arbiter and standard setter. As the years rolled by, however, many liberals became disenchanted with U.S. action as international policeman or bored with straight reporting and turned instead to the more sensational outpourings of the New Left.[3]

Newsweek paid tribute to Ascoli's "fervent and anti-Communist liberalism," which "made the magazine a lively force during the early '50s—more provocative than the weeklies such as *The New Republic* and *The Nation*, then languishing, and more forceful than the monthlies such as *The Atlantic* and *Harper's*, then largely literary." But it noted that later, "Ascoli found himself increasingly out of sympathy with the New Frontier as well as with the thaw in the cold war."[4] The *National Observer* noted that in the previous year *The Reporter*'s "most lucrative sources of advertising revenue, book clubs and publishers, cut back on their advertisements."[5]

Although financial problems loom large in press accounts of *The Reporter*'s closing, no former staff member believes that they were the real reason for its demise. The magazine was losing money, it is true; but it had always lost money. At the end, according to information *Newsweek* attributed to "insiders," it was losing about three-quarters of a million dollars annually. But these deficits were being made up, *Newsweek* also reported, by profits from the magazine's parent company, The Fortnightly Corporation, which had bought radio-TV station WBOY in Clarksburg, West Virginia, in 1963, and owned other properties.[6] And while it was true that *The Reporter*'s traditional advertisers—book clubs and publishers—had cut back, the magazine in its final years had acquired an impressive array of corporate advertisers: BOAC, Boeing, Buick, Columbia Broadcasting System, General Dynamics, General Electric, Hertz, IBM, Japan Air Lines, Western Electric, and Weyerhaeuser, among others.

Financial success, of course, had never been a primary aim of the magazine. Henry Luce, who always claimed idealistic motives for his publishing venture, admitted that in the early years of *Time* magazine "the bitch goddess sat in our outer office"—and it is just possible that she is still there.[7] Whoever may have sat in the outer office of *The Reporter*, it was not the bitch goddess of financial success. A professor, not a businessman, Ascoli could not be accused of being overly attentive to the business aspects of his magazine. In contrast to the editorial side of the magazine, there was considerable turnover on the business side. One former staff member says, "We had a series of advertising directors, each one with another idea, no one terribly consistent," and adds that the advertising staff was "incredibly small." Another, whose opinions may reflect some thwarted ambition and wishful thinking, argues that the magazine "never had the benefit of smart business publishing. That's plain *fact.* On the business side of the magazine it was a disaster." Ascoli, presumably, would reply that he was not interested in "smart business publishing" if that would interfere with the magazine's primary function of conveying facts and ideas. He wanted to put out a quality magazine that would help Americans exercise their responsibilities in the world. If publishing such a magazine meant incurring financial deficits, he and his wife were prepared to make them up. One former staff member with some experience of the business side of *The Reporter* recalls:

> Of course I learned about it from the perspective of the tax loss, which is an Alice-in-Wonderland way of looking at it. About a year after I left the magazine I ran into Ascoli on the street in New York and asked him how things were going. "Terrible," said Max. "You should see the taxes we had to pay this year." "Max," I said, "there are worse things than making a profit."

Even though it had never been an economic success, *The Reporter* was, in Ascoli's words, a succès d'estime: it had a reputation and influence out of all proportion to its circulation, and thus it was a valuable property. During its last two years Philip Horton, Shirley Katzander, and Donald Allan, among others, were actively looking around for sources of money for the magazine. Both the *Washington Post* and McGraw-Hill appeared to be interested in acquiring *The Reporter*, but not on terms that Ascoli felt he could accept. The problem was not one of money but of control: Ascoli was unwilling to let the magazine out of his hands. He had strong paternal feelings about it, and when asked about his reluctance to come to terms with prospective purchasers, he would reply, "Would you sell your daughter?" ("The response to that," says a disgruntled former staff member, "is, 'Well, would you murder her?' ") Wishful thinking persisted among many former staff members. They were intensely loyal to the magazine—or to their own individual conceptions of it, which may not be quite the same thing—and some of them refused to believe that it could not have been continued, if only Ascoli had not stood in the way. At the same time, their comments make it clear that the magazine would have had to change if it had continued. Here, of course, we get into an "iffy" area: it is hard to argue with people's dreams. But it is also hard to argue with their nightmares, and Ascoli had one that he described in 1971 as a kind of "mania": "Above all I could not help thinking of another magazine that had been every inch a one-man show—the *American Mercury*—and what it has become."[8]

H. L. Mencken's decision to leave his *American Mercury* in 1933 may have been prompted in part by the Depression, although the circulation and influence of the magazine were already on the decline by 1928.[9] The Depression seems to have been not so much the cause of his leaving as the occasion of his leaving. In the same way, the Vietnam War, often cited as one of the causes of *The Reporter*'s demise, seems to have been its occasion rather than its cause. Ascoli, approaching seventy, was simply getting too old to run the magazine the way he had run it in the past, and he had found no one that he was willing to name as his successor. His health was not good; he had had quite a serious operation in the summer of 1967. Many of those with whom he had felt most congenial had left the staff. Marya Mannes had departed in the summer of 1963 after a disagreement with Ascoli over the magazine's policies. Robert Bingham and Douglass Cater, the two young Harvard men on the original staff who had contributed so much to the magazine as managing editor and Washington editor, respectively, had both left in 1964, the former to go to the *New Yorker* and the latter to work for President Johnson. Gouverneur Paulding, one of Ascoli's closest friends on the staff, had died in 1965. Some of the others to whom he had felt close—Robert Gerdy, Harlan Cleveland, and William Lee Miller—had all been gone for some time. Those who had left had been replaced by very competent people, but it was not the same. The newer staff members tended to regard the magazine as a journalistic enterprise; Ascoli seems always to have regarded it more as a graduate seminar.

Ascoli's decision to close the magazine should not really have come as a great surprise to his staff. (One former staff member admitted later, "In retrospect, we were foolish not to see that the end was drawing near.") Another former staff member, who had known Ascoli before the magazine started, realized in the 1950s that Ascoli was never going to be able to turn *The Reporter* over to anyone else. Ascoli himself had apparently not yet reached this realization. As evidence that he was trying to establish *The Reporter* as a continuing institution, Ascoli in 1971 cited the three "alter egos"—his expression—that he brought in at various times: Harlan Cleveland, Irving Kristol, and Dwight Martin. Cleveland seemed the most promising: he stayed for two years as executive editor and another year as publisher (1953-56). Kristol stayed exactly one year as executive editor (1958-59), and Martin stayed only several months as coeditor (1963-64). None of them seemed to work out, and one former staff member has advanced a theory as to why they did not:

> He kept looking for the "alter ego," as I say. *I don't think he really wanted to find one.* If he had found somebody else like him, that would have meant that there was somebody else like him, wouldn't it? And I don't think that was what he wanted to find.

The comments of one of these "alter egos" tend to confirm this theory: "Ascoli hired me as his deputy, because he was sick. Then he got well, and he discovered that he didn't know what a deputy was."

A member of the original staff, describing some of the factors that led him to leave the magazine in the midsixties, says:

> I found myself disagreeing with the editorial policy more and more often, and this was strange, because in the past I had always been very enthusiastic about it. I guess another factor, I'd have to say, was my dawning awareness that the magazine probably wasn't being constructed in such a way as to go on forever. You know, *The Reporter* existed to do a certain job at a certain time, which may have been the same as presenting the ideas of Max Ascoli to the world. And then when that job was finished, there was no reason for it to exist.

By the midsixties Ascoli seems to have begun moving toward the same conclusion: "The idea of ending *The Reporter* had been growing on me for about three years. I kept telling myself, my wife, and a most faithful associate John Borghi [general manager of *The Reporter*] : I have to do it before my seventieth birthday."[10] In July 1967 he underwent a number of medical tests in preparation for a risky hip operation; his seventieth birthday—June 25, 1968—was less than a year away. In November 1967 he and Philip Horton finally came to a parting of the ways: Horton was kept on the masthead and the payroll for the next four months, but George Bailey was called in to take over some of his duties. The staff did not know precisely what had taken place between Ascoli and Horton; they did know that Horton would be leaving the magazine and was actively looking for a new position.[11]

Then came President Johnson's dramatic and unexpected announcement on March 31, 1968, of his withdrawal from the presidential race. This appears to have been the precipitating factor that led to Ascoli's finally making the decision that he had been moving toward for some time. It can hardly be labeled the "cause" of *The Reporter*'s ending. That ending, at least in Ascoli's mind, had become inevitable; the only question was when it would come. The President's surprising action—which Ascoli felt was a betrayal of critical but loyal supporters like himself—seems merely to have hastened the inevitable. Within a week and a half Ascoli made his own surprising announcement that *The Reporter* would close with the issue of June 13. And for those of its staff and readers who had assumed that the magazine would be an ongoing institution, the announcement came as a shock.

A number of readers, as well as the staff, were unwilling to see the magazine die without some effort to save it. One subscriber wrote to Ascoli:

> I am not a rich man by any means but would gladly pledge several hundred dollars a year to help keep *The Reporter* going. There must be many other long-time subscribers who feel as I do. You might not prefer to run a magazine that way, but damn it we need you.

Toward the end of April, Ascoli received a letter signed by Adolf Berle, Kenneth Crawford, John Dos Passos, Paul Douglas, Roscoe Drummond, Christopher Emmet, Henry Kissinger, Robert Murphy, William Proxmire, and others, asking to meet with him to see if there was some way of keeping *The Reporter* going. A few days later one of the initiators of the letter wrote again to Ascoli conceding that in the interim he and his friends had realized—apparently they had been talking to a number of others—that it was simply too late to do anything to save the magazine.

Having decided against continuing *The Reporter*, Ascoli arranged with *Harper's* for that magazine to take over *The Reporter*'s unexpired subscriptions. He also hoped that *Harper's* would provide an outlet for his occasional writings and those of some of his staff. He joined *Harper's* with the title of consulting editor and brought with him Meg Greenfield as Washington editor and Claire Sterling, Edmond Taylor, and George Bailey as foreign correspondents. This arrangement, as a number of *The Reporter*'s staff had predicted, was an extremely short-lived one; and of the five who went to *Harper's*, only Claire Sterling had anything published in that magazine. Ascoli's and Meg Greenfield's names appeared on the *Harper's* masthead for one issue only, that of July 1968. *Harper's* did not represent the same brand of liberalism as *The Reporter* had, and too great a cultural and generational gap existed between Ascoli, the liberal born in Ferrara, Italy, in 1898, and *Harper's* young editor, Willie Morris, the liberal born in Yazoo City, Mississippi, in 1934.[12] Almost as soon as he arrived at *Harper's*, where he expected to write a column called "Speaking for Myself," Ascoli learned that Morris planned to publish an article in the August issue on the same subject Ascoli was writing about—the conflict between the traditional liberals and the theorists of the New Left—but by another writer. That writer was Arthur Schlesinger, Jr., and the article was to be an expansion of the commencement address Schlesinger had just given at the City University of New York.[13] Ascoli was furious:

> I had not been consulted. . . . It seems that Willie found the speech so lovely and so endearing that he took it, although he knew that Max was working on the subject. I told him this week that I would write nothing for Harper's. . . . They have shocked me profoundly. I have been an editor for over 20 years and you cannot find in New York or in the whole international market a writer who has been treated that way by me. It can be justified only by an incurable youthfulness.[14]

Ascoli resigned at once from *Harper's*, and Meg Greenfield apparently left at the same time. The names of Claire Sterling, George Bailey, and Edmond Taylor, for more than a year and a half afterward, remained on the *Harper's* masthead. And then those names, too, disappeared.

One of the things that had discouraged Ascoli in the final months of *The Reporter* was the feeling that he had lost touch with his readers. Yet the more than three thousand letters he received after the announcement of the closing of the magazine revealed the hold he had on their affections. A woman in Logan, Utah, wrote simply: "I was a charter subscriber and have been a subscriber most of the time, and am now. It has been unique and I shall miss it." A historian and occasional contributor wrote to compare *The Reporter* to the *Nation* in its earliest years under E. L. Godkin. A former staff member wrote: "You have been true, over the years, to your own highest ideals; and the Reporter reflected it to the very end." A number of letters came from readers who disagreed with Ascoli's views on Vietnam. One wrote: "I am 51 years old, and the kind of liberalism I grew up with included tolerance, something today's dissenters take but do not give. I do not share your admiration for President Johnson, but neither do I agree with his belittlers." Another reader, noting that he did not subscribe to all of Ascoli's views on Vietnam, added that "your opinion was needed more and more as a reminder that liberals are just as prone to sheeplike behavior as any other variety of mammals."

One note that runs through many of the letters is gratitude for the effect *The Reporter* had on its readers' lives. One man wrote: "Your editorials and articles have always served to teach me—in a way that no one else has taught me—no one else, in my experience, except my father." A reader in Utica, New York, who was both a parish priest and a professor of sociology, wrote: "I think I have learned more about the in-depth workings of our democratic processes from 'The Reporter's Notes' and Max Ascoli's editorials than from any other source, extinct or extant."[15] Another academic, who had contributed several articles over the years, wrote to express heartfelt gratitude for the magazine, and added somberly that its closing seemed to be one of those historic events that mark the end of an era.

NOTES

1. Cf. Charles Kadushin, *The American Intellectual Elite* (Boston and Toronto: Little, Brown, 1974), p. 18: "Fifty years of the sociological study of organization can be summed up by saying that organizational goals are almost always subverted from their original purpose by the self-serving needs of organizational life."

2. Richard F. Shepard, "The Reporter, 19, Will Die June 13," *New York Times*, April 10, 1968, p. 56. The *Times* story contains some misleading information: its

list of contributors to *The Reporter*—James Thurber, Diana Trilling, Gore Vidal, Vladimir Nabokov, Ray Bradbury, and Dean Acheson—is so unrepresentative as to be ludicrous.

3. "Price of Consistency," *Time*, April 19, 1968, p. 60.

4. "The Reporter's End," *Newsweek*, April 22, 1968, pp. 66-67.

5. "Zest for Editing Fades, and Mr. Ascoli Calls a Halt to Reporter Magazine," *National Observer*, April 15, 1968, p. 14.

6. "The Reporter's End"; "Reporter Mag Buys Rustcraft's WBOY," *Radio-Television Daily,* March 12, 1963, pp. 1, 3.

7. Luce's comment is quoted in Robert T. Elson, *Time Inc.: The Intimate History of a Publishing Enterprise, 1923-1941* (New York: Atheneum, 1968), p. 82. The reference, of course, is to William James's famous complaint about "the moral flabbiness born of the exclusive worship of the bitch-goddess SUCCESS. That—with the squalid cash interpretation put on the word success—is our national disease."

8. The decline of H. L. Mencken's *American Mercury* from one of the most sparkling and influential magazines of the 1920s to an obscure, ultraconservative organ with miniscule circulation in the 1960s is related in Frank Luther Mott, *A History of American Magazines*, vol. 5: *Sketches of 21 Magazines, 1905-1930* (Cambridge, Mass.: Harvard University Press, 1968), pp. 2-26, and M. K. Singleton, *H. L. Mencken and the American Mercury Adventure* (Durham, N. C.: Duke University Press, 1962). The character of the *American Mercury* began to change markedly after Mencken left it in 1933. Lawrence Spivak, who owned the magazine from 1939 to 1950, wrote in 1959 that if he had known what was going to happen to the magazine after he sold it in 1950, he "would have buried it. . . . It is a shame that a magazine that contributed so much and earned a great name in its day, should come to its present low state." (Quoted in Mott, p. 26.)

9. Mott, pp. 17-18; Singleton, pp. 194, 199-201, 216.

10. Letter from Ascoli to the author, May 19, 1971.

11. In the fall of 1968 he went to work for the New School. See "New School Picks Program Chief," *New York Times,* September 4, 1968, p. 24.

12. Morris has given a good account of his intellectual development in his autobiography, *North Toward Home* (Boston: Houghton Mifflin, 1967). Less than three years after the dispute with Ascoli, Morris resigned from *Harper's* after a dispute with its publisher. See "Hang-Up at *Harper's*," *Time*, March 15, 1971, p. 41.

13. The article was "America 1968: The Politics of Violence," *Harper's*, August 1968, pp. 19-24.

14. Quoted in Henry Raymont, "Max Ascoli Leaves Harper's in Rift over Schlesinger Article," *New York Times*, June 14, 1968, p. 44.

15. *The Reporter*, June 13; 1968, p. 8.

Chapter Seven

Conclusion: End of an Era?

In the press, the networks, politics, the church, the schools and universities, and in commerce, the pressures today are running in favor of the conformist majority that offers the popular and easy answers to our problems.

James Reston, 1967

If a noble and civilized democracy is to subsist, the common citizen must be something of a saint and something of a hero.

George Santayana, 1905

Sometimes historical events seem to be timed so as to coincide neatly with our notions about historical movements. Anyone today rash enough to pass historical judgment on his own era might be tempted to pick 1968—the year not only of *The Reporter*'s demise but also of the withdrawal speech of President Johnson, the murders of Martin Luther King, Jr., and Robert Kennedy, the Chicago street riots during the Democratic Convention, and the election of Richard Nixon as President—as marking the end of the era of dominance of liberalism in America. At the end of the 1960s there was a feeling in the air that postwar American liberalism, at least as represented by such prominent liberal spokesmen as Arthur Schlesinger, Jr., John Kenneth Galbraith, and Hubert Humphrey, had reached a dead end.[1] American failures in Vietnam had focussed public attention on the shortcomings of liberalism: the Vietnam War was seen as a symptom of something seriously wrong with American society. A junior at Harvard gave this comment on Schlesinger's article, "Vietnam and the End of the Age of Superpowers," which had appeared in the March 1969 issue of *Harper's*:

> If the U.S. is not only to get out of the war in Vietnam (which Professor Schlesinger and other "liberal anti-Communists" turned against only because it was clear we were losing), but to insure that we will never again engage in crusades to destroy popular revolutions and the people themselves, we must recognize and repudiate the liberal ideology and policies which have made us the leading counter-revolutionary power in the world today.[2]

One way to document the "feeling in the air" at the end of the 1960s is to recall the titles of a few books that appeared at the time: Robert Paul Wolff's *The Poverty of Liberalism* (1968); Theodore Lowi's *The End of Liberalism* and Theodore Roszak's *The Making of a Counter-Culture* (1969); Zbigniew Brzezinski's *Between Two Ages*, Andrew Hacker's *The End of the American Era*, and Charles Reich's *The Greening of America* (1970). In their different ways all of them expressed the belief that the era of postwar American liberalism was coming to an end. And in the April 1971 issue of *Playboy* appeared Jack Newfield's article, "The Death of Liberalism." In the summer of 1976 Walter Dean Burnham of the Massachusetts Institute of Technology gave as his view that

> the American political system is gripped by a pervasive and intractable crisis. Quite a few writers, sensitive to its existence, have identified it with the "end of liberalism." Surely this is part of it: Carter's nomination will virtually certify that activist liberalism, as a national political force, is now in receivership. But the strange demise of liberalism is only part of a more general crisis of authority.[3]

It is ironic that the reputed death of liberalism should coincide with the death of *The Reporter*, the most widely read magazine that identified itself as "liberal." For some of the shortcomings in liberalism that were most instrumental in bringing about liberalism's crisis at the end of the sixties were the very shortcomings against which *The Reporter* had fought for nearly two decades. Four of these shortcomings are worth looking at briefly: the fallacy of pinning a label like "liberal" on someone and then expecting him to conform to one's own conception of what that label means; the failure to think through, in historical perspective and on a long-term basis, what America's role in the world should be; the fallacy of considering man as a totally political animal; and the failure of liberalism to define its own principles, particularly its central principle of freedom.

The fallacious idea that someone's total constellation of beliefs can be described by a simple label is a persistent one, but it rests on two unjustified assumptions: that such labels can be precisely defined in such a way as to have the same meaning for everyone, and that there are only as many possible patterns of belief as there are labels. Common labels like "liberal," "conservative," and "radical"—though popular and inevitable—are only abstractions. As accurate descriptions of real people they are artificial and inadequate. In many respects it is fitting to label Milton a "Puritan," but this label hardly prepares one for his views on sex and divorce. Dr. Johnson can confidently be described as a "conservative," but this label fails to explain his proposing a toast to a slave insurrection in the West Indies. Byron has indisputable claims to being regarded as a "Romantic," but those who think this label describes his views on all questions will be flabbergasted by his extravagant admiration for the neoclassicist Alexander Pope.

Likewise, a "liberal" does not always display what other liberals regard as a proper constellation of beliefs. When he escapes the pigeonhole to which he has been assigned, his independence can sometimes be quite upsetting. Norman Cousins has described the consternation among readers of the *Saturday Review* when John Lear, science editor of that predictably liberal magazine, expressed his doubts about the wisdom of water fluoridation:

> We were prepared for a storm but we had no idea it would assume the proportions it did. . . . The controversy on our letters pages grew week by week. We were accused of know-nothingism, of consorting with extremists and political rowdies. . . . We lost some readers we especially regretted losing.[4]

Similar consternation occurred among some readers of *The Reporter*, but over a longer period of time and over far more consequential matters. When Max Ascoli expressed his editorial doubts about such liberal heroes as Adlai Stevenson and John Kennedy, when he failed to cheer the liberal proclamations of the end of the Cold War, and when he supported the foreign policy of that uncouth Texan, Lyndon Johnson, his views did not conform to what some articulate liberals regarded as the proper pattern. To oppose Joe McCarthy for his attacks on the civil liberties of his countrymen in the 1950s was considered properly "liberal"; to oppose Ho Chi Minh for his attacks on the civil liberties of his countrymen in the 1960s, unfortunately, was not. And Ascoli, who believed he was consistently adhering to the same principles, was accused by some of turning "conservative."[5]

If by its very existence *The Reporter* revealed this flaw in liberal thought, it explicitly attacked a second shortcoming in liberalism—the failure to think through, in historical perspective, America's role in the world. In the 1950s, liberals supported American involvement in the world almost automatically. They regarded anyone who opposed foreign aid as a mindless conservative. But by the end of the 1960s liberals saw foreign aid as merely a portent of future military involvement and viewed it with skepticism or outright hostility. In 1969 Schlesinger, still a spokesman for many liberals, was saying explicitly what Ascoli had detected in his arguments during their 1956 debate: "The policy of total involvement in the world is incompatible with the policy of social reconstruction at home."[6]

Schlesinger's argument was carried to an absurd and depressing conclusion in 1970 by Andrew Hacker. Believing Americans to be no longer capable of making the sacrifices necessary to carry out their country's responsibilities as a world power, Hacker argued that the only realistic thing for America to do was to withdraw politically and militarily from the rest of the world and attempt to become another Sweden or Denmark.[7]

Nor all liberals agreed with Schlesinger or Hacker. Zbigniew Brzezinski, like Hacker a critic of liberalism and like him a contributor in 1960 of an article to *The Reporter*, expressed in 1970 the opposite conclusion that "a continental society like the United States could not survive by becoming merely another Sweden; it would not survive internationally and it is not even certain that it would find a satisfactory balance between domestic material needs and spiritual aspirations." Believing that no powerful nation—he cited Japan as an example—could safely remain isolationist, he maintained that America in particular, "the first global society," could not afford to evade its world responsibilities.[8] This is what *The Reporter* had been saying consistently for nineteen years.

The third liberal fallacy attacked by *The Reporter*—the inclination to see life in wholly political terms—was cited by Brzezinski as a fundamental cause of "the crisis of liberalism." He asserted that liberals had only begun to discover that *belief* is necessary to society, as they had, like the Communists, "underestimated the psychological and spiritual dimensions" of social well-being.[9] In the same year that Brzezinski's book appeared, Eugene Genovese, who would probably consider himself a radical liberal, was illustrating the persistence of this fallacy by arguing that America's "*spiritual* crisis can only be resolved through the creation of a *political* movement capable of realizing our national ideals by the reordering of our economic and social priorities through the restructuring of our economy."[10]

In his editorials in *The Reporter* Ascoli repeatedly attacked the fallacy that an essentially spiritual crisis could be solved by purely political means. His rejection of this fallacy lay at the heart of his objections to both Fascism and Communism. He stated his basic position clearly in 1949, the year *The Reporter* began:

> There is a limit to what politics can do in any organized community, national or international, however managed. It can certainly be used for the relief of human misery. It can give not only something else but something better—a betterment of actual and present suffering. But we cannot ask politics to do what even religion could not do: to give men eternal peace or eternal happiness.[11]

Alexis de Tocqueville, to whom Ascoli acknowledged his indebtedness, had a similar view of the limitations of politics. Political "solutions," Tocqueville believed, could lead only to bureaucratic pettiness and red tape:

> Men place the greatness of their unity in the means, God in the ends; hence this idea of greatness, as men conceive it, leads us to infinite littleness. To compel all men to follow the same course towards the same object is a human conception; to introduce infinite variety of action, but so combined that all of these acts lead in a thousand different ways to the accomplishment of one great design, is a divine conception.
>
> The human idea of unity is almost barren; the divine idea is infinitely fruitful. Men think they manifest their greatness by

> simplifying the means they use; but it is the purpose of God which is simple; his means are infinitely varied.[12]

The fourth fallacy in liberalism—the failure to define its principles, and especially its central principle of freedom—is perhaps the most crucial. It is part of the liberal's temperament to flee from precise formulation of his beliefs; but sometimes this flight reaches almost comic proportions, as is shown by the following comment of Walter Cronkite to an interviewer:

> I think of myself as a true liberal. And in my mind, a true liberal is someone who is not bound by doctrines or committed to a point of view in advance. Not even to a philosophical position in advance. He is one who examines each issue on its merits and makes his decision on that basis. . . . I am opposed to evil, I suppose, and this is just the end of my opposition.[13]

A man giving an interview, to paraphrase Dr. Johnson, is not upon his oath; but still one is tempted to ask how a person is able, without a doctrine or philosophical position, to determine rationally just what the "merits" of an issue are.[14]

Defining freedom is, of course, a necessary prelude to formulating a precise definition of liberal principles. But frequently liberals—and they are hardly unique in this respect—are content to let the word remain an abstraction. Then they can wave it like a banner or use it as a battle cry, happy that even if it may not carry a precise meaning it will at least bear a favorable connotation. Yet, like other abstractions, it embodies a multitude of meanings. Two of these, at least, are opposed in their tendencies: "Freedom *from something* is a great deal, yet not enough. It is much less than freedom *for something.*"[15]

The first of these meanings is embodied in the currently fashionable word "liberation"—liberation from the past, from outmoded ways of thinking, from restrictive customs, from traditional morality, and often from personal responsibility. Winnie Verloc in Conrad's *Secret Agent* finds this kind of freedom—at its worst—just after plunging the kitchen knife into her husband's breast: "She had become a free woman with a perfection of freedom which left her nothing to desire and absolutely nothing to do."[16]

The second of these meanings is illustrated by D. H. Lawrence's comment: "Men are not free when they are doing just what they like. The moment you can do just what you like, there is nothing you care about doing. Men are only free when they are doing what the deepest self likes." "It is never freedom," he maintains, "till you find something you really *positively want to be.*"[17] This second concept of freedom, which has an ethical or religious component, is part of the liberal heritage and is found throughout the writings of liberals like Milton and Jefferson. Distinguishing between "license"—roughly the seventeenth-century equivalent of "liberation"—and "libertie," Milton says that those who love the latter "must first be wise and good" (Sonnet XII). Jefferson, in "*A Summary View of the Rights of British America*, writes: "The God who gave us life, gave us liberty at the same time: the hand of force may destroy, but cannot disjoin them." And Tocqueville found this concept prevalent in the America he visited in the early 1830s: "The Americans combine the notions of Christianity and liberty so intimately in their minds that it is impossible to make them conceive the one without the other."[18]

One could hardly make such a statement about the beliefs of contemporary Americans. Yet one almost has to say this about Ascoli. His private definition of freedom is a startling one. Discussing the element common to the youth revolution in America and the protest movements associated with it, he said in early 1971:

> I would call it a revolt against Christ. Better to call it a revolt against freedom? Although it's an abstract word. But to use the word Christ is not proper for other reasons—it is too churchy.

Ascoli often felt that he couldn't articulate this belief publicly:

> Many of us know that man's destiny cannot possibly be decided by what we can encompass with our experience and measure with our reason but are nevertheless prevented from talking about God by fear of confessional disputes and by distaste for the still prevailing verbiage of religiosity. Sometimes it is easier to call a spade a spade than to recognize publicly our devotion to God.[19]

Constrained, then, to use the abstraction "freedom." Ascoli laid himself open to criticism for vagueness:

> In the nineteen years he published *The Reporter* . . . it was never clear that Max Ascoli ever understood what he meant by his favorite word, "liberalism." He was convinced that it had something to do with freedom but beyond that his personal definition became crusty and flaked around the edges.[20]

Yet in his annual Christmas editorials Ascoli was able to express, at least in part, his deepest beliefs about freedom, which were at the heart of his belief in liberalism. In his Christmas editorial for 1959, commenting on complaints by a number of political figures about America's "national and individual inadequacies," he wrote:

> But to overcome this sense of individual and collective failure—we respectfully suggest—there is no better way than finding refuge, each one of us *in interiore homini*, as Saint Augustine put it.
>
> This does not mean either to condone or fall in love with ourselves. It does mean to be quiet, at least once in a while, and try our utmost to find that safe nook where we can listen to the voice of our inner life. There is no human being in whom the capacity to hear this voice has been utterly and irreparably deafened. You can call it as you wish—voice of the spirit, or voice of God. But even in those miserable countries where all the powers of government are incessantly conditioning the thoughts, the actions, the instincts of men—even in the Communist countries the citizen cannot possibly be made to forget that there is something in him which belongs to another world and which the government cannot reach.[21]

If the modern world has indeed entered a "post-Christian" era, if increasing secularization is a tendency in human thought, and if a number of theologians find themselves constrained to assert that God is dead—then there was a belief at the heart of *The Reporter* that was antithetical to much that by the end of the 1960s had become intellectually fashionable. This antithesis was not widely noticed during most of the lifetime of *The Reporter*, partly because Ascoli did not parade his deepest beliefs, and partly because men acting out of differing philosophical beliefs can agree on many immediate political questions. But as American society became more sharply divided after the midsixties, and as Ascoli felt increasingly compelled to speak out in criticism of the "revolt against freedom," it became evident that *The Reporter* was out of step with many American intellectuals. Under such conditions *The Reporter* could be expected to thrive, or even to continue, only at the price of changing its character. But its character, Ascoli believed, was its reason for being. And so, logically, it had to come to an end. In this sense, the closing of *The Reporter* might be said to coincide with the passing of an era.

Eric Sevareid, who was in on the beginning of the magazine, wrote to Ascoli at its end: "I'm sorry the experiment is over, but you proved some things: some light was shed, some talents discovered, and a lot of people are the better for your great efforts over the years."[22] Disappointed as his staff and readers may have been at the closing, Ascoli himself kept some perspective by thinking of the precedent set by his father-in-law, Julius Rosenwald. Rosenwald, who seemed to believe that institutions, like human beings, had a natural lifespan, established a philanthropic institution, the Rosenwald Fund, with the unusual provision that both income and principal should be spent within twenty-five years of his death.[23] Ascoli, who married Marion after her father's death, has commented: "Yes, I actually thought of my father-in-law: twenty years of teaching, twenty years of editing, and as many years as God will give me dedicated to writing."[24]

Probably the best epitaph for *The Reporter* and for the efforts of its editor, staff, and contributors is the epitaph Ascoli wrote for his friend Dag Hammarskjöld: "What a good man has done in the full light of history never goes to waste. It always arouses ingenuity and goodness in other men."[25] Even those who disagreed with *The Reporter*'s stand on Vietnam or on other issues would probably agree that during its nineteen years it set a standard for journalistic excellence and integrity. American journalism can never have too many such examples.

What becomes of the liberal ideas expressed by *The Reporter* will be decided by the actions of those who call themselves liberals, for the survival and the success of freedom depend on the use men make of their freedom. The full story of *The Reporter* cannot be revealed by its history, but only by the subsequent lives of those it touched and changed—its editor, its staff and contributors, and perhaps most of all, its readers.

NOTES

1. For an extended discussion, pro and con, of the alleged death of liberalism at the end of the sixties, see William Gerber, *American Liberalism* (Boston: Twayne,

1975), pp. 13-40. Cf. Nicholas von Hoffman, *Left at the Post* (Chicago: Quadrangle Books, 1970), p. 209.

2. *Harper's*, May 1969, p. 6. Cf. Douglass Cater, *Dana: The Irrelevant Man* (New York: McGraw-Hill, 1970), p. 178.

3. Walter Dean Burnham, "Jimmy Carter and the Democratic Crisis," *New Republic*, July 3 and 10, 1976, p. 17.

4. Norman Cousins, *Present Tense* (New York: McGraw-Hill, 1967), pp. 57-59.

5. Ascoli would agree, I think, with Morris Cohen's statement in *The Meaning of Human History*, 2nd ed. (1947; reprinted LaSalle, Illinois: Open Court, 1961), p. 275: "Hatred of cruelties practiced by totalitarian states is not only perfectly consistent with liberalism but a necessary adjunct of any liberalism that can hope to survive in the world of today and tomorrow."

6. Arthur Schlesinger, Jr., "Vietnam and the End of the Age of Superpowers," *Harper's*, March 1969, p. 49.

7. Andrew Hacker, *The End of the American Era* (New York: Atheneum, 1970), pp. 228-29.

8. Zbigniew Brzezinski, *Between Two Ages* (New York: Viking, 1970), pp. 247, 298, 307-08.

9. Ibid., pp. 236-54.

10. Eugene Genovese, "A Massive Breakdown," *Newsweek*, July 6, 1970, p. 27. Emphasis supplied.

11. Ascoli, *The Power of Freedom* (New York: Farrar, Straus, 1949), p. 113. Cf. Daniel Patrick Moynihan, "Politics as the Art of the Impossible," in his *Coping: Essays on the Practice of Government* (New York: Random House, 1973), pp. 248-58; Peter Drucker, *The Age of Discontinuity* (New York and Evanston: Harper and Row, 1969), pp. 123, 248, 309; and William Lee Miller, *Of Thee, Nevertheless, I Sing: An Essay on American Political Values* (New York and London: Harcourt Brace Jovanovich, 1975), pp. 102, 109.

12. Alexis de Tocqueville, *Democracy in America*, ed. Phillips Bradley (New York: Vintage, 1958) vol. 2, pp. 386-87. This idea finds expression in literary works as varied as Emerson's "Brahma," Whitman's "Passage to India," Dostoevsky's *Brothers Karamazov*, and Robert Penn Warren's *All the King's Men.*

13. Oriana Fallaci, "What Does Walter Cronkite Really Think?" *Look*, November 17, 1970, p. 62.

14. This aspect of liberalism has frequently been criticized. See, e.g., Garry Wills, *Nixon Agonistes* (Boston: Houghton Mifflin, 1970), pp. 353-55; Louis Hartz, *The Liberal Tradition in America* (New York: Harcourt, Brace, and World, 1955), pp. 270-71, 305; and Eric Goldman, *Rendezvous with Destiny* (1952; rev. ed. New York: Random House, Vintage, 1956), pp. 240-42.

15. Czeslaw Milosz, *The Captive Mind* (1953; reprinted New York: Random House, Vintage, 1955), p. 34. Emphasis in the original.

16. Joseph Conrad, *The Secret Agent*, Chapter 11.

17. D. H. Lawrence, *Studies in Classic American Literature*, reprinted in *The Shock of Recognition*, ed. Edmund Wilson (New York: Modern Library, 1955), pp. 910-13. Emphasis in the original.

18. *Democracy in America*, vol. 1, p. 317.

19. Ascoli, "Christmas 1961," *The Reporter*, December 21, 1961, p. 14.

20. Robert Sherrill, "Weeklies and Weaklies," *Antioch Review*, Spring 1969, p. 35.

21. Ascoli, "Christmas, 1959," *The Reporter*, December 24, 1959, p. 4.

22. *The Reporter*, June 13, 1968, p. 8.

23. See Daniel J. Boorstin, "From Charity to Philanthropy," in his *The Decline of Radicalism* (New York: Random House, 1969), pp. 40-68.

24. Letter from Ascoli to the author, May 19, 1971.

25. Ascoli, "The Future of the U.N.," *The Reporter*, October 26, 1961, p. 23.

Selected Bibliography

Since sources are fully identified in the notes, this bibliography is confined to a listing of Max Ascoli's writings in English (and in one case, in French) outside the pages of *The Reporter*, and to a number of the most relevant books and articles by others. A useful bibliography of books and articles on liberalism in America can be found in William Gerber, *American Liberalism* (Boston: Twayne, 1975), pp. 269-300. Articles in *The Reporter* since the beginning of 1953 are indexed in *The Reader's Guide to Periodical Literature.*

Ascoli, Max. "After the Fascist Hoax." *Free World*, February 1946, pp. 22-25.

——. "Campus Riots and the U.S. Government." *Wall Street Journal*, May 27, 1969, p. 22.

——. "Coalition Government in Italy." *Free World*, June 1944, pp. 513-17.

——. "Communism in Italy." *Commonweal*, October 24, 1947, pp. 32-35.

——. Critique of Charner M. Perry, "The Arbitrary as Basis for Rational Morality." *International Journal of Ethics* 43 (January 1933): 154-57.

——. "Dulce et Decus pro Dictatore." *American Scholar* 6 (Summer 1937): 365-71.

——. "Education in Fascist Italy." *Social Research* 4 (September 1937): 338-47.

——. "Fascism, Atomic Power and UNO." *Free World*, March 1946, pp. 18-22.

——. "Fascism in the Making." *Atlantic*, November 1933, pp. 580-85.

——. "The Fascisti's March on Scholarship." *American Scholar* 7 (Winter 1938): 50-59.

——. "Freedom of Speech." *American Scholar* 9 (Winter 1939-40): 97-110.

——. *Georges Sorel.* Paris: Librairie Paul Delesalle, 1921.

——. "Government by Law." In *Political and Economic Democracy*, edited by Max Ascoli and Fritz Lehmann, pp. 229-42. New York: W. W. Norton, 1937.

——. *Intelligence in Politics.* New York: W. W. Norton, 1936.

——. "Italy, an Experiment in Reconstruction." *Annals of the American Academy* 234 (July 1944): 36-41.

——. "Land of the Free." *Survey Graphic*, April 1936, pp. 247-50.

——. "The Lesson of Italy." *Social Research* 11 (May 1944): 135-51.
——. "Mussolini in the War." *Yale Review* 29 (June 1940): 776-90.
——. "Neo-Fascism: Italian Sample." *Commonweal*, February 1, 1946, pp. 398-400.
——. "No. 38 Becomes a Citizen." *Atlantic*, February 1940, pp. 168-74.
——. "A Note of Dissent on 'Economics Today.' " *Social Research* 4 (May 1937): 203-08.
——. "Notes on Roosevelt's America." *Atlantic*, June 1934, pp. 654-64.
——. "Notes on San Francisco." *Free World*, January 1946, pp. 19-22.
——. "On the Italian Americans." *Common Ground* 3 (Autumn 1942): 45-49.
——. "On Mannheim's 'Ideology and Utopia.'" *Social Research* 5 (February 1938): 101-06.
——. "Political Parties." In *Political and Economic Democracy*, edited by Max Ascoli and Fritz Lehmann, pp. 205-16. New York: W.W. Norton, 1937.
——. "Political Reconstruction in Italy." *Journal of Politics* 8 (August 1946): 319-28.
——. "Politics in Italy." *Free World*, January 1946, pp. 19-22.
——. "Postscript on San Francisco." *Free World*, August 1945, pp. 13-17.
——. *The Power of Freedom.* New York: Farrar, Straus, 1949.
——. "The Press and the Universities in Italy." *Annals of the American Academy* 200 (November 1938): 235-53.
——. "Realism Versus the Constitution." *Social Research* 1 (May 1934): 169-84.
——. *Remembering Doctor McLean.* New York: New York Hospital-Cornell Medical Center, n.d. [1971].
——. "The Right to Work." *Social Research* 6 (May 1939): 255-68.
——. "The Roman Church and Political Action." *Foreign Affairs* 13 (April 1935): 441-52.
——. "The Rosselli Brothers." *Nation*, July 3, 1937, pp. 10-11.
——. "Society Through Pareto's Mind." *Social Research* 3 (February 1936): 78-79.
——. "The Test of San Francisco." *Commonweal*, June 15, 1945, p. 212.
——. "There Will Be Peace: An Editorial." *Free World*, June 1946, pp. 5-6.
——. "War Aims and America's Aims." *Social Research* 8 (September 1941): 267-82.
——. "What Total Diplomacy Means." *Nation*, May 20, 1950, pp. 489-90.
—— et al. "Round Table No. 28: World Government vs. United Nations." *Free World*, June 1946, pp. 22-29.
—— et al. "Round Table No. 28, Part II: World Government vs. United Nations." *Free World*, July-August 1946, pp. 21-26.
——. Review of *The New Party Politics*, by A. N. Holcombs. *Social Research* 1 (May 1934): 253-54.
——. Review of *The Intelligent Man's Review of Europe Today*, by G. D. H. and Margaret Cole. *Social Research* 1 (May 1934): 254-55.
——. Review of *The Method of Freedom*, by Walter Lippmann. *Social Research* 1 (August 1934): 389-92.
——. Review of *Public Opinion and World Politics*, by Quincy Wright. *Social Research* 1 (August 1934): 400-01.

——. Review of *The American Adventure*, by M. J. Bonn. *Social Research* 1 (August 1934): 401.
——. Review of *The Curse of Bigness*, by Louis D. Brandeis. *Social Research* 2 (February 1935): 122-23.
——. Review of *Thorstein Veblen and His America*, by Joseph Dorfman. *Social Research* 2 (August 1935): 391-93.
——. Review of *Law and the Lawyers*, by Edward Stevens Robinson. *Social Research* 3 (November 1936): 505-07.
——. Review of *The Political Philosophy of Hobbes*, by Leo Strauss. *Social Research* 4 (February 1937): 127-29.
——. Review of *The Good Society*, by Walter Lippmann. *Social Research* 5 (February 1938): 118-20.
——, ed. *The Fall of Mussolini: His Own Story by Benito Mussolini* [with a preface, "The Mussolini Story" by Max Ascoli, pp. 9-71]. New York: Farrar, Straus, 1948.
——, ed. *Our Times: The Best from The Reporter.* New York: Farrar, Straus and Cudahy, 1960.
——, ed. *The Reporter Reader.* Garden City, N.Y.: Doubleday, Anchor, 1956.
—— and Arthur Feiler. *Fascism for Whom?* New York: W. W. Norton, 1938.
—— and Fritz Lehmann, eds. *Political and Economic Democracy.* New York: W. W. Norton, 1937.
——. Introduction to *Navigating the Rapids 1918-1971: From the Papers of Adolf A. Berle,* edited by Beatrice Bishop Berle and Travis Beal Jacobs. New York: Harcourt Brace Jovanovich, 1973.
Bendiner, Robert. "What Kind of Liberal Are You? : A Classification of the Species." *Commentary*, September 1949, pp. 238-42.
Brock, Clifton. *Americans for Democratic Action: Its Role in National Politics.* Washington, D. C.: Public Affairs Press, 1962.
Brzezinski, Zbigniew. *Between Two Ages: America's Role in the Technetronic Era.* New York: Viking, 1970.
Cater, Douglass. *The Fourth Branch of Government.* New York: Random House, Vintage, 1959.
——. *Power in Washington: A Critical Look at Today's Struggle to Govern in the Nation's Capital.* Random House, Vintage, 1965.
Coates, Willson and White, Hayden V. *The Ordeal of Liberal Humanism: An Intellectual History of Western Europe*, vol. 2. New York: McGraw-Hill, 1970.
Cohen, Morris R. *The Faith of a Liberal: Selected Essays.* New York: Henry Holt, 1946.
Coles, Robert , ed. *The Geography of Faith: Conversations Between Daniel Berrigan When Underground, and Robert Coles.* Boston: Beacon Press, 1971.
Delzell, Charles F. *Mussolini's Enemies: The Italian Anti-Fascist Resistance.* Princeton: Princeton University Press, 1961.
Drucker, Peter. *The Age of Discontinuity: Guidelines to Our Changing Society.* New York and Evanston: Harper and Row, 1969.

Ellul, Jacques. *The Technological Society*. New York: Random House, Vintage, 1964.

Felker, Clay S. "Life Cycles in the Age of Magazines." *Antioch Review*, Spring 1969, pp. 7-13.

Fermi, Laura. *Illustrious Immigrants: The Intellectual Migration from Europe 1930-41*. Chicago and London: University of Chicago Press, 1968.

Forcey, Charles. *The Crossroads of Liberalism: Croly, Weyl, Lippmann, and the Progressive Era, 1900-1925*. New York: Oxford University Press, 1961.

Gerber, William. *American Liberalism*. Boston: Twayne, 1975.

Goldman, Eric. *The Crucial Decade—and After: America: 1945-1960*. New York: Random House, Vintage, 1960.

——. *The Tragedy of Lyndon Johnson*. New York: Knopf, 1969.

Gregor, A. James. *The Fascist Persuasion in Radical Politics*. Princeton: Princeton University Press, 1974.

Halberstam, David. *The Best and the Brightest*. New York: Random House, 1972.

Hamby, Alonzo. *Beyond the New Deal: Harry S. Truman and American Liberalism*. New York: Columbia University Press, 1973.

Hartz, Louis. *The Liberal Tradition in America*. New York: Harcourt, Brace and World, 1955.

Horton, Philip. *Hart Crane: The Life of an American Poet*. New York: Norton, 1937.

Kadushin, Charles. *The American Intellectual Elite*. Boston and Toronto: Little, Brown, 1974.

Kissinger, Henry A. *American Foreign Policy: Three Essays*. New York: W. W. Norton, 1967.

Lekachman, Robert. "Political Magazines." *Book Week*, November 17, 1963, pp. 24-28.

"Liberal Anti-Communism Revisited: A Symposium." *Commentary*, September 1967, pp. 31-80.

Lipset, Seymour M. *Political Man: The Social Bases of Politics*. Garden City, N.Y.: Doubleday, 1960.

Mannes, Marya. *Out of My Time*. Garden City, N. Y.: Doubleday, 1971.

"Max Ascoli." *Current Biography*, February 1954, pp. 7-9.

Miller, William Lee. *Of Thee, Nevertheless, I Sing: An Essay on American Political Values*. New York: Harcourt Brace Jovanovich, 1975.

Morris, Willie. *North Toward Home*. Boston: Houghton Mifflin, 1967.

Mott, Frank Luther. *A History of American Magazines*, vol. 5: *Sketches of 21 Magazines, 1905-1930*. Cambridge, Mass.: Harvard University Press, 1968.

Osgood, Robert E. *Ideals and Self-Interest in America's Foreign Relations: The Great Transformation of the 20th Century*. Chicago: University of Chicago Press, 1953.

—— et al. *America and the World: From the Truman Doctrine to Vietnam*. Baltimore and London: Johns Hopkins Press, 1970.

Peterson, Theodore. *Magazines in the Twentieth Century*. 2nd ed. Urbana, Ill. : University of Illinois Press, 1964.

Reston, James. *The Artillery of the Press*. New York: Harper and Row, 1967.

Rivers, William L. *The Opinionmakers.* Boston: Beacon Press, 1967.

Roosevelt, James, ed. *The Liberal Papers.* Garden City, N. Y.: Doubleday, Anchor, 1962.

Rovere, Richard. *Senator Joe McCarthy.* New York: Harcourt Brace, 1959.

Schapiro, J. Salwyn. *Liberalism: Its Meaning and History.* Princeton: D. Van Nostrand, 1958.

Schlesinger, Arthur M., Jr. *The Vital Center: The Politics of Freedom.* Boston: Houghton Mifflin, 1949.

Sherrill, Robert. "Weeklies and Weaklies." *Antioch Review*, Spring 1969, pp. 25-42.

Singleton, M. K. *H. L. Mencken and the American Mercury Adventure.* Durham, N. C.: Duke University Press, 1962.

Sinyavsky, André [Abram Tertz]. *The Trial Begins and On Socialist Realism.* New York: Random House, Vintage, 1960.

Sterling, Claire. *The Masaryk Case.* New York: Harper and Row, 1969.

Taylor, Edmond. *Awakening from History.* Boston: Gambit, 1969.

Tocqueville, Alexis de. *Democracy in America.* Edited by Phillips Bradley. 2 vols. New York: Random House, Vintage, 1958.

Trilling, Diana. "On the Steps of the Low Library: Liberalism and the Revolution of the Young." *Commentary*, November 1968, pp. 29-55.

Trilling, Lionel. *The Liberal Imagination.* New York: Viking, 1950.

Vietnam: Why: A Collection of Reports and Comments from THE REPORTER. New York: The Reporter Magazine Company, 1966.

"What Is a Liberal—Who Is a Conservative? A Symposium." *Commentary*, September 1976, pp. 31-113.

White, Theodore H. *The Making of the President—1964.* New York: Atheneum, 1965.

——. *The Making of the President—1968.* New York: Atheneum, 1969.

Wills, Garry. *Nixon Agonistes: The Crisis of the Self-Made Man.* Boston: Houghton Mifflin, 1970.

Index

ABOUT THE AUTHOR

Martin K. Doudna, a Washington civil servant during the Kennedy and Johnson administrations, is now an associate professor of English at Hilo College, Hilo, Hawaii, where he has worked extensively on the relationships between literature and the history of ideas. He is presently preparing articles on Whitman, Thoreau, and Orestes Brownson. His play *Have You Any Room for Us?* was published in 1975.